THIS BOOK IS
DESIGNED TO BE
USER FRIENDLY

Clear, Fast Understanding

- Complete, accurate information.
- Clear, consistent organization,
  with many headings and subheadings.
- Bullet lists, rather than long sentences and paragraphs.
- Numerous examples illustrate every principle.
- Bold italic type used to emphasize key points.

Instant Reference

- Short Contents on outside front cover.
- Complete Contents on inside back cover and facing page.
- Complete Index at end of book.
- Summary information on outside back cover.
- Main headings start on new page.
- Headings in large, bold italic type.
- Page numbers in large, bold italic type,
  on outside top corner of each page.
- Cross-references give page numbers,
  as well as section headings.

Easy Reading

- Simple, clear, concise language.
- Simple, short words, sentences, and paragraphs.
- Personal tone and active tense.
- Ragged-right lines, without hyphenation,
  make for easier, faster reading.
- Clean type style, of easily readable size.

Easy Use

- Flexible covers.
- Light in weight. Can easily be carried in a briefcase.

TO YOU, THE USER OF THIS BOOK

DEDICATION

I dedicate this book with love and gratitude to my Mother, Thelma Bolsky, who passed on, on November 11, 1984, Veterans Day. She was a real veteran who loved God and all of God's creation. Her whole life was devoted to loving and helping people. I pray and know that she still does so.

I also dedicate this book with love and gratitude to my Father, David Bolsky, a fine person who passed on many years ago.

"The Lord bless you and keep you.
May God's face shine upon you and be gracious to you.
May God look upon you kindly and give you peace."
(Blessing of Saint Francis of Assisi)

WHO THIS BOOK IS FOR

This book will be useful in college and industry courses on writing.

This book will be useful to management, technical, and professional people, as a comprehensive reference handbook. These people, to whom writing is important, but who usually devote a small part of their time to writing, may wish to skim over this book ... then to study in detail sections that interest them ... and finally, to refer to this book as needed.

Technicians, secretaries, clerical people, and others do some writing. These people may also want to skim over this book, study in detail sections that interest them, and refer to this book as needed.

Those who devote most of their time to writing, may want to read all of this book.

This book has many ideas that can be useful to people in their personal writing, to help make letters, notes, etc., clear and specific.

TERMINOLOGY USED IN THIS BOOK

Users, Not Just Readers

Your document, or whatever it is that you are writing, will be *used* by people, not read for entertainment. That's why I use the term *users* in this book, rather than "readers."

These users are your customers — whether they are associates in the office next door, or consumers in the marketplace. Whoever they are, how well you do on the job depends on your pleasing them. So ... stay close to your customers. Make sure that you know who they are, what they *need* and *want*, and how you can best give it to them.

Documents

The principles in this book apply to any kind of informative material, both technical and non-technical — articles, books, booklets, documents, handbooks, letters, manuals, memos, papers, plans, proposals, reports, resumes, and speeches. For the sake of consistency, I use the term *documents* in this book.

Better Examples Shown First

This book contains many examples. One reviewer suggested that examples might have more impact if the poorer way was shown first, and then the better way. However, I show the *better way first*. I believe that the way a person sees *first*, is the way that they are likely to remember the best.

6

ABOUT THE WRITING STYLE OF THIS BOOK

My writing style differs from the *sentence-and-paragraph* style that is usually taught in school, and that is used in most writing. That style is fine for "literature" type material — fiction ... and non-fiction articles, books, etc., intended to entertain or to inform lightly, rather than to inform in detail or to instruct.

My style is *heading-list-and-example* based. That is —
- *Headings* — Instead of several paragraphs or even pages without headings, I use many headings and subheadings, so as to organize the material as much as possible.
- *Lists* — Instead of presenting information in sentences and paragraphs, I use bulleted lists as much as possible.
- *Examples* — Instead of writing at length about the ideas that I wish to convey, I write about them as briefly as possible and then I *illustrate* them as much as possible.

I believe that heading-list-and-example style is more effective for informing users in detail, and for instructing them. This style —
- Lends itself to better organization.
- Is more clear and precise.
- Requires fewer words.
- Takes less time to read and refer to.

Of course, many people are used to, and may prefer the sentence-and-paragraph style, and there are good books on that style. One reviewer of this book felt that that style might be easier to read, perhaps especially for beginners. I disagree. However, I'm sure that there's a place for both styles. (See *Pleasantness* on page 16.)

The writing style that I both present and use in this book is similar to that which I used in my three books on computers. Prentice-Hall, the publisher, has received favorable comments on the writing style of these books, as well as on the content, and I have also. In case you are interested in these books, they are **The C Programmer's Handbook, The VI User's Handbook,** and **The UNIX®️ System User's Handbook (UNIX** is a trademark of AT&T).

Typical of the comments received about these books by Prentice-Hall, and by myself about these books and about other material that I have written at AT&T, are the following (I present these comments just to show that users like this style) — "It is everything a handbook should be. Short, clear, easy to use, to the point, and complete." ... "Excellent reference material. Very well organized, very well structured. Well put together. Well thought out." "It is the closest thing to a real handbook that I have ever seen."

THIS BOOK CONTAINS PRINCIPLES, NOT FORMAL RULES

One of the reviewers of the draft of this book said, "Parts of the book contradict advice given. For instance, on page 15, titles are not parallel (violating advice given on page 87)." It's true. The paragraph title "First And Foremost" on page 15, isn't parallel with the following titles in that section — "Integrity," "Warmth," "Organization," etc. However, I feel that the point that I make in the "First And Foremost" paragraph should be emphasized with that heading, and that making the emphasis is more important than is making the titles parallel.

That reviewer, and others, may disagree with me. They may feel that making the titles parallel is more important. But I do have what I consider to be a good reason for not being parallel.

The reason that I bring this up, is to ask you to regard the material in this book as *principles* for good writing, not as formal rules. Don't feel that you must follow every principle, exactly as it's given. Adapt the principles to your writing style, and to what the users of your document need and want.

COMPUTER TOOLS FOR WRITING

If you do your writing on a computer ... or if you have access to a computer and you have the skill and the time, or if you have a typist to enter your document into a computer ... then you may want to find out about computer programs that aid writers. There are spelling and thesaurus programs. There are also programs to check grammar, punctuation, word and sentence length, writing style, etc. One program that does much of this is the *UNIX*® *System WRITER'S WORKBENCH*® *Program*, developed by AT&T.

SUGGESTIONS FOR USING THIS BOOK

You can use this book as a text for study on your own ... as a text in a course on writing ... and/or as a reference book. I'll list each section of the book here, and briefly tell you what it's about.

However, before listing the sections, I'll first discuss the aids to using this book.
- *Brief Contents* on the outside front cover.
- *Complete Contents* starting on the inside back cover and facing page.
- *Index* starting on page 187.
- *Basics* section on the outside back cover, with a very brief summary of basic principles for better writing.

Note — Read through the following three sections, if you want to.

TO YOU THE USER — Contains introductory information.

BETTER WRITING — Also introductory in nature. Briefly discusses what better writing is about and why you may want to bother to write better.

WRITING — Discusses the steps in writing — Clarifying ... Developing ... Preparing ... etc. You may be especially interested in the Starting section, which has dozens of ideas for countering "writer's block" ... when you don't feel like writing.

Note — The rest of this book contains reference type information. You can read through these sections, or just skim through them so as to become familiar with them, and later refer to the information and the examples as needed.

DESIGN — PRINCIPLES — Discusses what I consider to be *the* basic principles for better writing — Accuracy ... Clarity ... Completeness ... etc.

DESIGN — BASIC ASPECTS — Discusses the types of basic support information that many documents need — Cross-References ... Footnotes ... Format ... etc.

DESIGN — BASIC SECTIONS — Discusses the types of basic sections that many documents need — Covers ... Contents ... Index ... etc. It's arranged in the order that these sections would normally appear in a document.

DESIGN — BASIC AIDS — Discusses the types of aids to understanding that many documents need — Examples ... Illustrations ... Lists ... etc.

DESIGN — PHYSICAL — Discusses the Importance Of The Physical Design Of A Document ... Color ... Computer Vs. Paper Documents ... etc. It's a subject that isn't often covered in books on writing, but that I feel is important to know about, as it opens up different options to you.

PRODUCTION — Discusses Typing ... Proofreading ... Printing ... and Distribution of a document. Perhaps you can leave these matters to others. Even so, I think it's important for someone who writes a lot to know about these matters.

FAIRNESS — Discusses, in detail, how to avoid sexism in writing. It's summarized and adapted from a Prentice-Hall guide for authors.

GRAMMAR — Discusses and illustrates basics such as Modifiers ... Dangling Modifiers ... Misplaced Modifiers ... etc. This and the following two sections are summarized and adapted from U. S. Government publications.

PUNCTUATION — Discusses the mechanics of Acronyms ... Apostrophes ... Brackets ... etc.

ACTION VERBS/ABSTRACT NOUNS — Lists action verbs, and the corresponding abstract nouns which writers often use instead of the action verbs. The action ***verbs*** are easier to understand.

ACKNOWLEDGEMENTS

My sincere thanks to all who have helped and encouraged me, including, but certainly not limited to —

Terry Alger, Larry Frase, Carolyn Nilson, Carl Simone, Bill Taylor, and Chris Yuhas for their useful recommendations.

Diane Anderson, who skillfully typeset this book. I also want to thank her for the other books and publications that she has typeset for me. She has always been highly professional, conscientious, and a pleasure to work with.

Larry Bernstein, Dan Clayton, and Nick Ippolito, managers at AT&T Bell Laboratories.

David Korn and Valerie Torres, good friends of mine.

Karen Olson, who helped and encouraged me at the time of my Mother's passing on, and who is helping so many people in her work with the homeless.

John Wait, my editor at Prentice-Hall.

My Search For God Study Group.

You, the user of this book. This book represents my philosophy of communicating by the written word. Thank you for reading this book and thus giving me the opportunity to share this philosophy with you. I hope that you find it of value.

PERSONS AND ORGANIZATIONS QUOTED

Also, I want to acknowledge the following persons and organizations, whom I have quoted in this book. The page numbers on which the quotes appear, are shown in the Index starting on page 187.

- American Press Institute
- Carolyn Boccella Bagin
- Business Week Magazine
- Winston Churchill
- Document Design Center
- Jo Van Doren
- Robert D. Eagleson
- *FORTUNE* Magazine
- Joseph M. Fox
- Harold S. Geneen
- Laura Grace Hunter
- John F. Kennedy
- Henry Wadsworth Longfellow
- W. Somerset Maugham
- George Orwell
- Blaise Pascal
- Norman Vincent Peale
- Prentice-Hall College Division
- Will Rogers
- David A. Schell
- James R. Squire
- Mark Twain
- U. S. Government Printing Office
- *The Wall Street Journal*
- C. A. Warren
- Westinghouse Electric Corporation
- Jan V. White
- Frank Lloyd Wright

BETTER WRITING

WHY WRITE AT ALL?

Often we are tempted just to tell something to people, rather than taking time to put it in writing. But there may be good reasons to put it in writing, even though it takes time and effort to do so.

This section discusses reasons for writing, when you have the choice of writing or of just talking to the persons involved. Often, of course, you don't have such a choice, as when you have to communicate formally to persons higher in your organization than you are, or to customers.

- *Reputation* — Better writing is important to do your job more successfully, and to enhance your reputation as an expert on the topic and as a good communicator. It is thus a stepping stone to higher level positions. Written material, whether it be a short memo or a long document, is very *visible*. Several, perhaps many people read what you write. If you write poorly, it reflects poorly, not just on your writing ability, but on *you*. And if you write in an outstanding way, then this reflects well on you in every way.

- *Schedules* — It may be difficult to get everyone together who needs to know the information. A document may help avoid having to call a meeting. Or, by preparing people with the information, a document may cut down the number of meetings that need to be held, and their length.

- *Accuracy And Self-Instruction* — You are likely to spend more time clarifying and organizing your thoughts, getting more information, checking on details, etc. if you write a document, than if you just tell the information to people. This extra effort will increase your knowledge of the topic and also the knowledge of the users of your document.

- *Organization* — Your writing reflects your thinking. If your writing is disorganized, it likely means that your thinking about the topic is disorganized. By taking time to organize what you write about a topic, you are really *investing* your time in organizing your thinking about it. Equally important, you should make the organization highly visible, with many headings and subheadings. "The University of California recently established a special faculty committee ... to study the prevalence of substandard writing among undergraduates. ... Its most conclusive finding was that difficulties in organization and structure predominate in poor writing; proficiency in the techniques of grammatical usage seemed a corollary of general ability to organize material logically. ... In short, good writers tend to be those concerned with organization, with the problems involved in forming relationships. The poor writers are those concerned primarily with mechanics. Moreover, poor writers, unlike the good, are totally unable to recognize good writing in others. One wonders whether those of us who have concentrated on spelling, punctuation, and vocabulary in our classes — that is to say, concentrated on writing rather than composition — have contributed to this disability in any way." (James R. Squire, "Tension on the Rope — English 1961")

- *Simplicity* — Use the simplest language possible. Never mind that you may be writing for PhDs. They need and want simple writing just as much as anyone else does.

- *Conciseness* — Here I refer, not to cutting out a word here and there, but to combining or eliminating duplicate and overlapping sections. This is another aspect of organization ... of clarifying and simplifying your thoughts.

- *Clarity* — Only after you have organized your writing, should you start clarifying and simplifying the paragraphs, sentences, and words.

- *Pleasantness* — One reviewer felt that sentence-and-paragraph style may be more pleasant to read. But I feel that heading-list-and-example style is clearer and quicker to read and to understand ... and thus, more pleasant. Perhaps most documents should consist of material in both types of styles.

BETTER WRITING IS IMPORTANT

> "The minute a thing is long and
> complicated it confuses. Whoever
> wrote the Ten Commandments made
> 'em short. They may not always be
> kept, but they can be understood."
> (Will Rogers. Weekly Article, March
> 17, 1935. Will Rogers Memorial,
> P. O. Box 157, Claremore, Oklahoma
> 74018)

In 1979, **FORTUNE** magazine talked to many successful corporate executives about what business schools should teach. Interviewers asked, "What kind of academic program best prepares business school students to succeed in their careers?" Executive after executive said, in frustration — *Teach them to write better*.

The importance of well organized, clear, and simple writing is infinitely more important today than it was years ago. The fate of many people, and even of the nation and of civilization itself depends on the operators of enormously complex systems in medicine, aviation, nuclear power, and the military.

Users often have difficulty understanding a document because of its *style* (or lack thereof!), rather than because of its *content*. This causes problems. The work takes longer. Misunderstandings occur. Perplexed users turn to co-workers for help, wasting everyone's time. As a last resort, users guess at what they should do, causing error, waste, and delay. Frustrated users may feel inadequate and lose confidence in themselves and their work.

One survey of document users found that they had the following complaints —
• Information spread over several documents.
• Hard to find information.
• Unneeded, incomplete, unclear, inaccurate, out-of-date information.
• Unrealistic, unclear, or nonexistent examples.
• Jargon.
• Poor or nonexistent Contents and Indexes.
• Confusing section numbering systems.
• Documents hard to order, and delivery took too long.

BETTER WRITING PAYS

A report by Robert D. Eagleson, the Australian government's Special Adviser on Plain English Matters, offers examples of the economic as well as the social benefits of plain English. The following is from his report, as presented in *Simply Stated*, February 1986 (Document Design Center, American Institutes for Research, 1055 Thomas Jefferson Street, NW, Washington, DC 20007). Figures are in U.S. dollar equivalents, although Eagleson gave many of the numbers in British pounds.

Note — This book is about writing in general, not about forms design. However, forms design is a form of writing. So I believe that these examples of savings that are possible, are applicable to any form of writing. The fifth example on the next page is about regular written material.

In Great Britain

- Fifty thousand airplane passengers fill out a form every year for the Customs and Excise Department to claim lost baggage. The form had an error rate of 55%. Redesigning the form reduced the error rate to 3%, saving staff 3,700 hours in processing. It cost the Department about $3,500 to rewrite the form but saved about $45,500 a year in processing costs.

- For the Department of Defense, civilians fill out 750,000 claims forms every year for traveling expenses. By rewriting the form, the Department cut the error rate by 50%, reduced the time needed to fill out the form by 10%, and cut the time to process it by 15%. This new form only cost about $16,500 to produce but saved 80,000 staff-hours or about $552,000 a year.

- The Property Service Agency improved the design of a contract work order and saved about $221,000 a year in typists' time.

- In 1984, the Department of Health and Social Security introduced plain English application forms for legal aid. The Department spent about $34,500 to develop and test the forms but saved about $2,069,000 in staff time for the year.

- In 1983, the UK Home Office produced a new application form for naturalization that takes people 15 minutes less to complete than the old one and, thus, saves the public 20,000 hours a year.

- The Passport Office reduced the time it takes to fill out the application form for one of its visas from 40 to 20 minutes on the average — significant because English is the second language of most of its users.

- The UK Land Registry introduced a new B10A enquiry about bankruptcy and showed an 80% decrease in the number of written questions and a 30% decrease in the number of complaints.

- The Office of Fair Trading produced a single form in straightforward language that replaced four forms that were previously required.

In the United States

- When the Federal Communications Commission (FCC) issued its regulations for citizen-band radio that were written in legalese, the agency needed five full-time staff members to answer the public's questions. After FCC rewrote the regulations in plain English, the questions stopped and the five staff members were assigned to other areas.

- Citibank revised its forms so that both its staff and its customers understand them. This reduced the time spent training staff by 50% and improved the accuracy of the information that staff gave to customers.

20

THE HIDDEN ISSUES IN BETTER WRITING

Like everything else, better writing has a price.

- It takes more time and effort to write a well organized, concise, and clear five-page document, than it does to dash off the same information in twenty muddled pages. "I have made this letter longer than usual because I lack the time to make it shorter." (Blaise Pascal)

- After you have spent the time and effort to write a clear five-page document, users may think — "Well, that's fine, but of course this topic isn't too deep." Whereas if you dash off twenty muddled pages, users may think — "Wow! This person is brilliant to be able to write about such a difficult topic."

If one is fortunate enough to have managers who want ... allow time for ... recognize ... appreciate ... and reward better writing, there's no problem.

WHAT IT TAKES
TO WRITE BETTER DOCUMENTS

- *Talent* — I believe that this book can help you to write *better* documents than you are presently writing. But as for really good ones, there's no substitute for talent. Just as it takes talent to be a top manager, a top musician, a top artist, a top scientist, etc. However, study can help you to do better, even though you may not have the talent to become a top expert.

- *Supportive Management* — I discuss this in *THE HIDDEN ISSUES IN BETTER WRITING*, on the preceding page. Your management must *really* want better writing, and better work in general ... not just pay lip service to it.

- *Good Role Models* — In the section on *PREPARING* on page 26, I suggest that you build up a collection of good documents that you feel are especially well written and well designed, that you can get ideas from. And, of course, if there are good writers around who are willing to spend some time reviewing your material, they can be extremely helpful to you, not just for what you are writing now, but for your writing and working skills in general.

- *Experience* — Just as there's no substitute for talent, so there's no substitute for experience. The more you do something, and do your best to do it *better*, the better you get at it.

- *Hard Work* — Recently I read a memo describing a block-type technique of writing. I thought that the technique was useful. But the following statement was, in my opinion, not very useful — "Less time is spent on writing. Because most blocks contain no more than a few sentences, the author has only to fill out the outline to come up with a completed document." This made it sound as if the outlining and the writing have been reduced to a mere routine, requiring little time, thought, or hard work. Which is nonsense. If you want to do a good job at anything, it requires much time, thought, and hard work.

WRITING

CLARIFYING

- *Define Your Purpose* — Talk to your manager, and to potential users of what you are writing and to their managers. Discuss orientation, content, level of detail, format, physical form, schedule, approvals. Perhaps also hold a design review at which you present your plans for the document, and at which interested persons can offer their opinions. Be sure to clarify just what users *need* and *want*. Do this clarification periodically.

  Remember that job titles (e.g., computer programmer, engineer, manager) tell you little about the people who will use your document. First of all, job titles may vary widely in different companies, or even in different departments of the same company. And second, people with the same job title may have very different educations, experience levels, abilities, interests, etc. Talk to many potential users to get a feel for who they are and what they need and want.

- *Respect Users* — Involve users as much as possible in your work. You may think that you know what they need. But they know what they want. Discuss the purpose with them until both they and you are clear as to what is *needed and wanted*.

24

- *Clarify If Your Purpose Is To —*
  - *Document* information for the record.
  - *Report* on what was done, or what was found.
  - *Discuss* the background and philosophy of the subject, so that others can decide what to do.
  - *Inform* people in a general way about the subject, just for their information.
  - *Recommend* what should be done.
  - *Instruct* people on (what, when, where, why, how) they should do something.
  - *Persuade* (advertise) people to buy, or do, or use something.
  - *Order* people as to (what, when, where, why, how) they should do something.

DEVELOPING

- *Research* — "Take what is best from the past, and build on that." Some or even much of what you need to write, may already have been written. Spend time researching material that you can adapt, or even incorporate into what you are writing.
  - Don't feel that you must do everything anew, reinventing the wheel. That may be called for in a *school* environment, where the goal is to test your knowledge and ability. But in a *business* environment, the goal is to do the job as effectively as possible.
  - Don't take the "Not Invented Here" attitude ... that if something wasn't done or written by you or by someone in your department, it couldn't be very good. Of course, if you use someone else's work, get permission if needed, and acknowledge their work.

- *Talk With Experts* — *People* may be more up-to-date than are documents. They can tell you on which points you should focus, and they can refer you to other people and to specific parts of documents that you need to read. And, of course, you can ask questions of people.

- *Talk With Users* — Talk with potential users, of all levels of experience and ability.
  - Experienced and capable users can provide you with detailed information and handy "tricks" that they have learned. On the other hand, they may know the material so well that they gloss over or downplay the importance of some information that may be important to less experienced or less capable users.
  - Less experienced or less capable users may call your attention to information needed by them.
  - New users may call your attention to the information and procedures that they need.

- *Observe* — Watch and talk with potential users. Make it clear to them that your purpose is to analyze their needs, not to evaluate them personally.

- *Participate* — Perform the task yourself, under actual conditions. "You soon discover, in a fire control director on top of a rolling destroyer or in the tight quarters of a submarine, how you would design equipment [or documents] differently ... if you had had this field experience earlier." (C. A. Warren)

PREPARING

- **Write Concurrently With The Project** — Don't think of writing as a necessary but boring postscript to a project. Think of it as a challenging part of the project, to be done concurrently with the project.

- **Seek To Be Creative** — Don't feel that writing a document is dull and uncreative. Of course, it isn't the same as writing a science fiction fantasy ... at least hopefully it's not, for the sake of the people who will have to use your document! But writing a well organized, concise, and clear document can be ... in fact, should be ... highly creative.

- **But Don't Be Creative In An Artificial and Harmful Way** — The author of one brochure was "creative" by using ten or more different colors, making nearly every page a different size, and using other such gimmicks. Aside from its costing a fortune to print, users complained that the brochure just wasn't well written. Here the author was creative in the sense of designing a document that was probably unique in its form and thus pleased the author, but that didn't please users by meeting their needs.

- **Build Up A Collection of Good Documents** — Start a collection of documents, booklets, foldout cards, etc. that you feel are especially well written and well designed. Then, when you need to write something, look through your collection for ideas. Spread out on a table all of the appropriate material. Consider —
  - What type of information should your document have?
  - What level of detail?
  - What organization?
  - What format?
  - What physical form?

- **Find Out What Services and Facilities Are Available** — Familiarize yourself with the services of Editing, Word Processing, Drafting, Art, and Printing. Do this before you start writing, so that you will know what facilities are available and what help you can get.

- *Give Reminders* — At all stages in writing, you will be working with people — subject matter experts, managers, editors, typists, artists, printers, etc. Always remember that — *Nothing is likely to happen on schedule without well-timed, diplomatic reminders*. To them, you are just one of many "customers" ... but to you, you are *The Customer*. So don't wait until a due date to contact them. Talk to them *before* the due date, and ask —
  - If they know what to do.
  - If they have any questions.
  - If they have, understand, and can follow your instructions and schedule.
  - (But don't rush them when there is no rush. It isn't fair to them. And it isn't fair to yourself either, because when you really do have a rush job, they may not pay attention.)

STARTING

"Writer's block" is a common phenomenon. This section offers suggestions on how to overcome it.

- *Clarify What You Are Supposed To Be Doing* — A frequent cause of difficulty in getting started, or in continuing with what you are doing, is not being very clear about what you are supposed to be doing. Talk with your manager, with associates, and with potential users. Then write a Purpose Statement, and check it with them.

- *Think Of Who, What, When, Where, Why, How* — Write out and answer these questions about your document.

- *Speak Up If You Aren't Getting Needed Support* — If the people who are supposed to be providing you with information or other support aren't doing so, or are doing so in a halfhearted manner, talk with your manager.

- *Face The Fear Of Criticism* — Perhaps you are afraid that your manager won't like what you are writing. Show him or her what you have written, and get feedback.

- *Brainstorm* — Do this yourself or with others. Write down all the ideas that you can think of, about what to put in the document and how to say it. Observe this rule — No evaluation (especially, no criticism) allowed during the brainstorming session.

- *Discuss Your Project With Associates* — This is an excellent way to clarify your thinking. Even if they know nothing about your project, sympathetic listeners can serve as a "sounding board" for your ideas. You are likely to get new ideas, especially if they ask questions about what you are doing. And, of course, if they do know about the project, they may provide useful ideas, information, and leads to other people or to documents.

- *Keep A List Of Points To Cover* — As you start to write, or before, ideas will come to you about what points to cover and what to say. Jot down these points on 3x5 cards, 8½x11 sheets, or in a notebook. Keep adding to, and organizing your list of points and your list of what to say. Gradually you'll see how to organize the material ... what points need to be added, deleted, or combined ... and how to say it.

- **Don't Feel That You Must Outline Before You Start** — You can, of course, start with an outline. But you don't have to. If you want, you can start with an outline of just a small part of the document. Or you can just start writing, with no outline at all.

- **Warm Up** — Sharpen your pencils. Make a phone call. Clean up your desk. Etc. But don't do this for too long. Get to work!

- **Start At Any Point In The Document** — You don't have to start at the beginning and continue to the end. You can start at any point that you feel like.

- **Organize The Document Into Small Sections** — Then think of only one at a time. Forget the rest.

- **Look At Past Documents On The Subject That You Are Writing About** — Look at examples of already written documents, by yourself or others, of the type of document that you are writing now.

- **Look At Your Own Past Documents That You Like, On Any Subject** — This will boost your confidence, and may give you ideas.

- **Read Something, On Any Subject, By An Author Whom You Admire** — You may be inspired by it.

- **Read Over The Parts Of Your Document That You Have Already Written**

- **Write In A Simpler Way** — Consider if you are trying to write in an official, complex way, either to impress users or because you think that is the way you are supposed to write. Try to write in a simpler way, as if you were speaking to users.

- **Set Aside A Daily Time Period** — Force yourself not to do anything else during this period, except work on your document. Even if you don't write a single word, don't do anything else — no phone calls, no putting your desk in order, no "To Do" lists, etc. It's **write**, or nothing.

- **Set A Quota For The Day And Then Stop** — Set a quota of how many pages you will write for the day. Then stop and do something else.

- *Put Your Feet Up* — Write at a table, with a hassock or chair on the other side of it so you can put up your feet.

- *Turn Off The Telephone*

- *Play A Radio, Or A Cassette Recorder* — Isaac Asimov is said to work with a radio blaring.

- *Use A White Noise Machine* — This produces sounds to mask distracting noise.

- *Line The Room With Cork For Silence* — Marcel Proust is said to have done this.

- *Splash Cold Water On Your Face*

- *Eat Something*

- *Take A Few Slow, Deep Breaths*

- *Exercise* — Stretch. Move your head forward and backward, and then sideways a few times. Rotate your arms. Walk around the building, or go outside for a walk.

- *Meditate* — Sit or lie quietly on the floor for a few minutes. Think of a relaxing scene.

- *Write In A Noisy Place* — Jane Austin is said to have done her writing in the middle of the family hubbub.

Note — Some reviewers thought that the following items were a bit "heavy," beyond the bounds of this book. But I wanted to make this section as complete as possible. So I've left these items in. They may help some people.

- *Get Medical Help* — If you consistently have a problem in getting started and continuing with your work, you may have a medical problem — low thyroid, allergies, etc.

- *Get Environmental Help* — There may be something in the environment that is affecting you — gas leak, allergy to some nearby equipment, poor air supply, pollution from inside or outside the building, etc.

- *Look At Your Living Habits* — Are you getting enough sleep? Are you eating a good diet? Are you overworking? Do you have a serious family or personal problem for which you may need help? Etc.

- *Consider If You Are In The Right Job* — Maybe you just aren't cut out for this sort of work.

EDITING

- *Sleep On It* — After you have your first rough draft, put it away for a day or two. You can then come back to it with a fresh viewpoint. If what you are writing is long and will take many days, do this periodically.

- *Edit And Re-Edit* — Also ask others, especially some people who will have to use what you write, to edit the material, or at least to comment on it.
  - Read your writing aloud, especially the key parts. Do this alone, and/or with someone else.
  - Write or type drafts double or triple space to allow room for changes.
  - When your draft becomes unreadable because of too many changes, get it retyped. A word processor makes this easier, because only the changes need to be entered.
  - If you find it hard to cut out material, then after you have cut it out, file it in an Archives File. So you won't feel that it's lost forever. This isn't just a psychological trick. You may indeed decide to put it back in the current document, or in a future document.

- *Organize* — Eventually, perhaps after you have finished writing, you should organize your material. Check it for the following —
  - Does it focus on your objectives, and not wander off into side issues?
  - Is every important point included?
  - Are main points emphasized?
  - Are subpoints placed under the correct main points?
  - Is the sequence logical?

TESTING

- *Don't Work Alone In A Vacuum* — Don't spend much time writing and heavily editing a "masterpiece" before you show it to anyone ... only then to find that it isn't what they need or want. Issue a first, rough draft soon. Send it to your manager, to some experts on the topic, and to some key potential users. Then talk with them, to ensure that you are proceeding correctly. Do this periodically.

- *Mark It As A Draft* — Write *DRAFT* on the front page (or on every page) of drafts, so people won't think that it's a final copy.

- *Put On Your Name* — Write your name/room/phone on the front page of drafts.

- *Put On The Date* — Date each draft on the first page, or on every page if you or others may take the pages apart.

- *Have A Master Copy* — With each successive draft, write all changes from everyone's copies (the changes that you choose to accept) on a master copy. Then discard the old drafts, so as to avoid confusion. Or file the old drafts in a corner of your bookcase, where they won't get mixed up with current drafts.

- *Ask For Comments In Red* — Ask people to use a red (or some other color that contrasts with the print) marker pen for comments. Maybe tape one to each review copy.

- *Send Photocopies* — When a good draft is ready, send photocopies to some or all potential users, so that they can use it at once, and also so that they can give you feedback as to its accuracy, completeness, clarity, etc.

- *Talk To People About Your Document* —
  - Don't just ask them if the meaning is clear to them. They may assume that they know what you mean. But they may be wrong. Ask them to tell you, in their own words, what they think it means. For procedures, ask them to actually carry out the procedures while you observe them.
  - Don't just do this with a couple of your friends, who probably are familiar with the topic and with the way you write. Do this with a representative sampling of the people who will be using your document, both experienced and inexperienced.

- **Field Test Your Document** — If your document is part of a system that will be undergoing field test, then test the document as part of the field test.

- **Study Use Of Your Document** — Consider doing a study such as the following. This was done for the user manual for a computer text editing system. The project manager hired several typists from a temporary employment agency, gave them the user manual and a computer terminal dialed into the system, and asked them to study the manual and learn how to use the system. The manager or an associate sat by each typist all the time and took notes, but didn't help them in any way. The typists were also asked to give a running commentary on what they were thinking and doing, as they studied the manual and used the system.

  The manager said that much useful information was obtained in this way that probably couldn't have been obtained in any other way. The manager also said that regular employees of the company wouldn't have wanted someone to watch over their shoulder, or even if they were ordered to participate, they might not have been totally frank in their comments. They might have thought that it was a test of them, rather than of the user manual. The temporary typists were hired specifically for the study, and thus they had no reason to fear that they were being tested since they didn't work for the company.

  One important finding is that the typists tended to blame themselves, rather than the user manual or the user interface of the system, when they couldn't understand something or did something wrong. So when people say that a document is "Okay," it may mean little or nothing.

Note — The following is from *Simply Stated* (May-June 1985).

"The Document Design Center announces a new service to help businesses improve consumer communications — a Usability Test Laboratory for documents.

"The lab is designed to let you and our evaluation specialists observe people as they use computer manuals, tutorials, or instruction booklets for VCRs or major appliances. It is also designed for focus-group sessions.

"A one-way mirror between the lab's observation room and the evaluation room allows unobtrusive viewing of people using your materials. State-of-the-art video and audio equipment record the test situation.

"DDC's evaluation specialists are experimental and cognitive psychologists who can help by:
1. analyzing the document's purpose and the characteristics of the users
2. developing test scenarios and setting criteria for selecting test subjects
3. observing and recording subjects' actions
4. identifying problems and working with DDC's writing and design specialists to improve the document's organization, style, and graphics

"For more information, contact the Usability Test Laboratory, American Institutes For Research, 1055 Thomas Jefferson Street, NW, Washington, DC 20007; (202) 342-5000."

- *Be Wary Of Readability Formulas* — Readability formulas that test word and sentence length, and other aspects of your document, can be useful. However, the main test of a document is, *Is it clear to the intended user?* The following is from a brochure of the *Document Design Center* (see address above). "We do not test documents with readability formulas. Research has shown that readability formulas don't measure how understandable or useful a document is. They don't show where a document needs fixing or how to fix it. The only useful way to test a document is to have appropriate samples of readers try to use it."

FOLLOWING UP

There are several ways to follow up on your document after it is issued —

- *Reader Reply Form* — This is the cheapest way. And probably the poorest, since you are likely to get little response. However, since it is so cheap, you might want to do it. At least put an address in the Introduction of your document, to which people are asked to send comments. Consider having a statement such as this — "Every user is responsible for the accuracy and up-to-dateness of this document. This means that anyone who finds an error, an unworkable idea, or anything that should be changed, should contact the author. No one is free to pass the buck and say that it's not his or her responsibility."

- *Questionnaires* — Send questionnaires, some time after the document is distributed, to all or to selected users.

- *Telephone Surveys* — Phone users, and ask for their general comments. Also ask specific questions.

- *Personal Interviews* — Make appointments with some users, and talk with them about the document. Again, ask specific questions.

- *Personal Observation* — Sit next to people while they are actually using your document, and observe how they use it. This assumes that your document is intended for use while one is performing some task.

DESIGN — PRINCIPLES

ACCURACY

- Don't rely on documents alone for source information. Personally check with experts.

- Doublecheck your final draft with source information and with experts. Even if you originally obtained the information from the same experts, check the final draft with them also. You may have misinterpreted or omitted something, or put something in the wrong context. Or, after reading your draft, the expert may realize that he or she was wrong, or didn't explain it correctly. Or the information may have changed.

- Proofread all drafts. Perhaps ask someone else to proofread them also. Even if your document was done on a word processor, you should check everything each time before you discard the old draft. Even if you made just a few changes here and there. Both typists and computers have been known to make mistakes.

- Ask someone to read your draft from the standpoint of — **Does it make sense?** They may catch obvious errors that you missed, because you are so familiar with the material that you didn't notice the errors.

- Make sure that examples are correct. One study of computer programming textbooks found that the textbooks had many wrong examples. Sometimes the examples violated the very rules that they were meant to illustrate.

- Don't leave out important details because they might "confuse" users.
  - "Harvey Brooks, a physicist who teaches government at Harvard, says much science coverage [in newspapers and magazines] is sound, but that most journalists have a 'low tolerance for ambiguity' and often leave out important details that might weaken a science story's conclusions...."
  (Reprinted by permission of **The Wall Street Journal**, © Dow Jones & Company, Inc. 1985. All rights reserved.)

CLARITY

- Make your document **be** clear and simple to use. Also make it **appear** so at first glance, so that users will want to use it! Which of the following sentences would you prefer to read?
  - **Say** "This won't work if it's bent."
    Not "A proper functioning of this component is critically dependent upon its maintaining dimensional integrity."
  - **Say** "Last year we changed the way we work."
    Not "During the preceding year the company was able to change the method of working."
  - **Say** "Our action was based on the belief that the company was making money."
    Not "The action was predicated on the assumption that the company was operating at a financial gain."
  - **Say** "They agreed to the contract."
    Not "They acceded to the proposition to approve the contractual relationship."

- Use ***short***, ***simple*** words, sentences, paragraphs, sections. Only one idea to a sentence. Only one group of related ideas to a paragraph. Only one subject to a section. "Short words are best and old words, when short, are best of all." (Winston Churchill)

Say	*Not*
about	approximately
also	additionally
although	despite the fact that
aware	cognizant
because	due to the fact that
begin	initiate
best	optimum
buy	purchase
change	modification
consider	take into consideration
do	accomplish
do	implement
failure	malfunction
finish	finalize
first	initial
for, to	for the purpose of
go	proceed
good	meaningful
help	facilitate
if	in the event that
improve	ameliorate
many	a large number of
much	a great deal of
name	designation
no	negative
now	at the present time
part	component
plan	conceptualize
problems	operational difficulties
scarce	in short supply
show	demonstrate
study	make a study of
use	utilize
yes	affirmative

- A survey of the American Press Institute (11690 Sunrise Valley Drive, Reston, Virginia 22091) indicated the following relationship between the number of words in sentences, and reader comprehension.

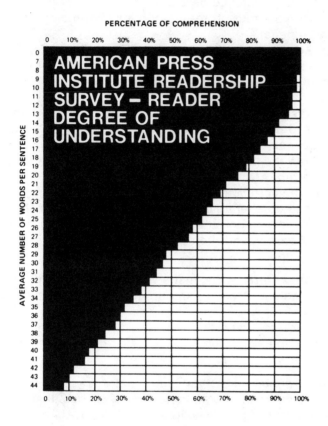

PERCENTAGE OF COMPREHENSION

- *Avoid meaningless modifiers* such as "actually," "definitely," "practically," "really," "very," "virtually." (But sometimes these words do emphasize your meaning.)

44

- *Avoid conjunctions* such as "and," "because," "but," "for," and "however" to start a new clause. Instead, start a new sentence. The second sentence may not be complete (as in the examples below). But so what? The main thing is that the sentence be *clear*.
  - *Say* "Take your foot off the gas pedal. *Then* step on the brake pedal."
    Not "Take your foot off the gas pedal *and then* step on the brake pedal." (*Note* — There is disagreement in the professional community about which form is clearer.)
  - *Say* "There isn't enough time available for the average manager to do everything that needs to be done. *Thus* it is necessary to determine the essentials and do them first. *Then* the manager can spend the rest of the time on things that are fun to do."
    Not "There isn't enough time available for the average manager to do everything that might be done, *and thus* it is necessary to determine the essentials and do them first, *and then* spend the rest of the time on things that are fun to do."

- In *conditional statements*, make words like "and," "if," and "or" stand out. Use uppercase and/or bold italic type.
  - *Say* "*IF* you are over 21,
    AND you live in the city,
    THEN you are eligible for ...."

- *Use active tense*. One study found that it takes up to 25% more time to understand passive sentences, than it does to understand active sentences. Also, users are more likely to misunderstand the passive sentences.
  - *Say* "Pat *did* the job."
    Not "The job *was done* by Pat."
  - *Say* "Our purchasing department *submitted* bids."
    Not "Bids *were submitted* by our purchasing department."
  - *Say* "We *discussed* the plan."
    Not "The plan *has been discussed* by us."
  - *Say* "Always *wear* safety belts."
    Not "Safety belts must always *be worn*."
  - *Say* "*Complete* the form."
    Not "The user *should complete* the form."
  - *Say* "Our staff *will plan* the meeting."
    Not "The meeting *will be planned* by our staff."

- Don't switch from active to passive tense in the same sentence.
  - *Say* "They **considered** the report but **took** no action."
    Not "They **considered** the report but no action **was taken**."
  - *Say* "**Turn off** power. **Disconnect** wires.
    Not "**Turn off** power. Wires **must be disconnected**.

- However, sometimes passive sentences are preferable to active ones.
  - "The call was heard by everyone" (passive) is preferable to "Everyone heard the call" (active) if you want to emphasize the word "call."

- Sentences that start with "It" and "There" can usually be simplified and shortened —
  - *Say* "We believe ...."
    Not "It is believed by us ...."
  - *Say* "They need not ...."
    Not "It is not necessary that they ...."
  - *Say* "Please do this today."
    Not "It would be appreciated if you would do this today."
  - *Say* "I recommend that ...."
    Not "It is my recommendation that ...."
  - *Say* "No changes have been made."
    Not "There have been no changes made."
  - *Say* "No uniform procedure is followed."
    Not "There is no uniform procedure followed."
  - *Say* "The group has a rule that ...."
    Not "There is a rule in this group that ...."
  - *Say* "The group has 6 people."
    Not "There are 6 people in the group."

- Using an *infinitive* ("to study," "to improve," etc.), is usually shorter and clearer than is using a phrase ("for the purpose of," etc.).
  - *Say* "Pat went home **to study**.
    Not "Pat went home **for the purpose of studying**."
  - *Say* "They increased the lighting **to improve** working conditions."
    Not "They increased the lighting **so that they could improve** the working conditions."

CONCISENESS

- *The Fewer The Words, The Better*. But —
  - Don't be so brief that your writing is difficult to understand, or may easily be misunderstood. "Telegraphic" style usually should be avoided.
  - Even though several short sentences may result in more words than one long sentence, often the several short sentences will be clearer.

- Present only *essential* information. Too much information can be just as bad as too little. It may delay publication of the document, obscure the essential information, confuse users, and unnecessarily consume their time. If you need to give little-needed information, make it brief. Or put it in an appendix, or in a supplement sent only to people who need or want it.

- Don't go into much detail on how hard it was to get the information for the document, do the study for it, write it, etc. You may want to ensure that your management knows how hard a time you had in doing the work. But others usually don't need or want to know this. Tell it to your management in a separate memo, or at the back of your document. But if you really feel that it's important for users to know, then tell it to them up front, perhaps in the *INTRODUCTION*.

- Just because you have spent much time and effort getting much information for your document, don't feel that you *must* include it all because "it's there" ... or that you *should* include it all to show how much work you've done. Include what users need and want. No more.
  - The producer of the Jedi movies told how over 100 people worked for a year to build a set, for a scene that took just two minutes in one of the movies. He said that in his opinion the biggest mistake made by producers of science fiction movies, is that when they spend much money to build a set, they then feel that they should get their money's worth by using the set as much as possible in the movie. He said that he feels they end up with the reverse of "getting their money's worth" ... they degrade the movie by using a set more than it should be used. He said that he felt that his set, on which 100 people worked for a year, was essential to the plot, but that two minutes was enough. More would have been too much.

- Don't repeat the same information in more than one place in a document, unless you have a good reason.
  - If you must do so, use the **same wording**, the **same order**, and the **same format** in each place ... unless you have a good reason for doing otherwise.
  - To totally eliminate any doubt, state that you are repeating the information, perhaps with different wording, in a different order, and with a different format. And tell the reason for the repetition and for the differences in wording, order, and format.

- Information to be used **just once** and in **just one place** (for instance, how to get a computer job number, which would only be done once at a computer center input counter), might not be needed in a document. In the document, you might just advise the user to ask for assistance at the time and place that it's needed. The reason that I say this, is because directions that perhaps can be told and demonstrated in just a minute or two, might take hours to write and even then not be clear. However, if the directions are critical and there's the possibility that no one might be around when the user needs them, then of course you should include the information in the document.

- Don't assume that the bigger a document is, or that the more documents there are, the better. Often the reverse is true.
  - "The project kept verging on a pitfall, and occasionally falling into it, that is very typical of large projects — that in our emphasis on documentation, we kept missing the mark and kept getting quantity rather than quality. We got stacks of documents a foot thick. It was clear that nobody was going to read them, and in fact nobody ever could have read them. They were useless." (From a debriefing report by the author of this handbook.)

- *Know When To Stop.*
  - A manager issued a memo asking people to take care of the new office furniture. The manager ended the memo with, "I earnestly request your cooperation with respect to the foregoing." "I earnestly request your cooperation" would have made a strong ending. But "with respect to the foregoing" spoiled it with a meaningless, bureaucratic phrase.

CONCRETENESS

- "In prose, the worst thing one can do with words is to surrender to them. When you think of a concrete object, you think wordlessly, and then, if you want to describe the thing you have been visualizing, you probably hunt about till you find the exact words that seem to fit it. When you think of something abstract you are more inclined to use words from the start, and unless you make a conscious effort to prevent it, the existing dialect will come rushing in and do the job for you, at the expense of blurring or even changing your meaning. Probably it is better to put off using words as long as possible and get one's meaning clear as one can through pictures or sensations." (George Orwell)

- Ideas are better and faster understood and remembered if they evoke *images* that are *concrete*, *meaningful*, and *familiar*. Relate ideas to conditions that users are likely to be familiar with in their everyday lives.
  - Saying that it's 239,000 miles to the Moon doesn't evoke much of an image. Saying that it's over nine times the Earth's circumference at its equator is better. Saying that it's about 80 times the one-way distance from New York to California is still better.

- Words that arouse an image in the user's mind are better than abstract words. Does the word refer to a *person*, *place*, or *thing* that can be *seen*, *heard*, *felt*, *tasted*, or *smelled?*
  - *Say* "It rained *every day* for a *week*."
    Not "A *period* of rainy weather set in."
  - Winston Churchill talked of *blood*, *sweat*, and *tears* ... not of courage, energy, and sadness.
Image	*No Image*
Fire	Combustion
Transistor	Device
Oscilloscope	Equipment
Light	Illumination

- *Specific language* conveys more information than does general language. Specific language is also more interesting. Thus it is more likely to be read, and also to be understood and remembered. Use specific names of people, places, and things, specific times, and specific actions.
  - *Say* "*Pat Smith* will go to the *ABC Company* in *New York* on *June 7*, to repair their *broken X1 machine*." *Not* "The *repairperson* will go to the *customer's premises* at the *earliest possible date* to work on the *subject equipment*."
  - *Say* "Visit our *book distribution center* and see our *customer service capabilities* for yourself." *Not* "Take our *familiarization tour*."
  - *Say* "Attached is our *7-page, 4-color brochure*." *Not* "Attached is *collateral material*."

- Minimize readers' need for —
  - *Calculations* — By providing detailed, specific conversion tables, graphs, charts, etc.
  - *Judgment* — By providing detailed, specific decision tables.
  - *Writing* — By providing a form on which users can simply check off or write in their responses.
  - *Interpretations* — By trying to consider all of the possible situations that may be encountered by users.

CONSISTENCY

- Be consistent in your use of **technical terms**. Pick the term that is most familiar to users, and that is short, simple, and specific, and stick to it. If there are several possible meanings to a technical term, stick to one and make clear which meaning you are using.
  - Don't say "machine" in one sentence, and then "computer," "processor," and "XYZ Computer" in others, when you always mean the same thing.
  - Don't refer to the same thing as a "command," an "instruction," and a "function" in different places.

- Be consistent in your use of **non-technical terms** and words. Using different words for the same thing gives variety, but it also causes misunderstanding.
  - Don't say "memorandum" in one sentence, and then "document," "report," and "paper" in others, when you always mean the same thing.
  - The following sentence has four non-matching pairs. "A large increase was reported in seven of the areas, and a less significant rise occurred in twelve instances." (Large, less significant ... increase, rise ... was reported, occurred ... areas, instances)

- Be consistent in your use of **verbs**. One can "press," "depress," "hit," "tap," or "strike" the keys on a computer terminal. And one can "enter," "input," or "type" data. Always use the **same word** for the **same action**.

- Be consistent in the **form of terms**. Are a "filename" and a "file name" the same thing? A "database" and a "data base?" "On-line" and "online?"

- Be consistent in **capitalization**. Are the "BEGIN," the "Begin," and the "begin" computer commands all the same thing? Always write it exactly as people will have to use it. If, for example, people have to enter a computer command name in all lowercase letters, then always write that name in all lowercase even if it is the first word in a sentence. This is important, not just for the sake of consistency, but to avoid confusing users and to **reinforce the correct behavior**.
  - **Say** "**begin** is the command to use when ...."
    Not "**Begin** is the command to use when ...."

- Be consistent in the *order of terms in sentences*.
  - *Say* "The BEGIN, IF, DOWHILE, and END commands are important. Use the BEGIN, IF, DOWHILE, and END commands ...."
  Not "The BEGIN, IF, DOWHILE, and END commands are important. Use the DOWHILE, IF, END, and BEGIN commands ...."

- Use *consistent language forms* for parallel ideas —
  - *Say* "*Turn* dial to 90. *Raise* the lever."
  Not "*Turn* dial to 90. Lever *should be raised*."

- Be consistent in *lists*.
  - *Say* "1. *Plan* before you write.
     2. *Select* what to include.
     3. *Organize* the material."
  Not "1. *Plan* before you write.
     2. *Selecting* what to include.
     3. *You should organize* the material."

- Suppose you have several pairs of lists, where you list good and bad ways of doing things. Don't put the good ways first in some lists, and the bad ways first in others. Or, if good to bad items are listed in the same list, don't start with the good in some lists, and with the bad in other lists. *Note* — I think that it's better to put the good or correct way *first*, because what people see first is what they may remember the best. They may not even bother to look at the second way at all, or they may just hastily glance at it.

- Be consistent in your use of *personal pronouns*. Don't say "I" in one place and then "we" in another. State at the start of the document how you will refer to everyone, and stick to it. *Note* — I suggest that you say *I* when you refer to yourself (the author). Saying "we" sounds pretentious and can be confusing, and saying "the author" also sounds pretentious.

- Be consistent in your use of *words*. Don't use the same word in two different senses in the same sentence or paragraph, or even in the same document.
  - *Say* "The *number* of items was modified *several* times."
  Not "The *number* of items was modified a *number* of times." (Better yet, in this case be specific and say, "The number of items was modified *ten* times.")

- Be consistent in your use of **conventions**.
  - Don't use a convention in your document that is very different from that used in your company or industry.
  - Don't use the same convention in two different ways, in the same document. For instance, in one document italic type was used in the **text** to denote computer command names, and in **examples** to denote variable information.

- Be consistent in **level of detail**. Don't go into great detail on topics with which you are very familiar, while you skim over or omit other, equally important topics. This may burden users with too much information on some topics, and deprive them of information on other topics. It may also confuse users, as they may naturally assume that the topics on which you have provided much information are the most important.

- Be consistent in **level of accuracy**. If it isn't possible to discuss everything at the same level of accuracy, clearly indicate the level of accuracy of each section.

- Check for **internal consistency** within a section or a document. If the same information is repeated in different sections or documents, perhaps in different formats, ensure that the information in one section or document doesn't contradict that in another. The possibility, or even probability, of this happening, is a good reason for minimizing duplication.

- Be consistent in your use of **examples**, **pictures**, **diagrams**, etc. Ensure that the text doesn't **say** one thing, while the example **shows** something different.

- It's especially important to pay attention to consistency when **updating material**. A common error is to update the material in one place, but not in another place of the same document. Another common error is to update the text but not the examples, or vice-versa. These common errors are good reasons for consolidating related information in one place.

- Consistency is important but, as with any good thing, it can be overdone.
  - Try to be consistent within the **present document**. If several persons are each writing separate sections of the same document, or separate documents that are related to one another, then give extra attention to ensure that what they write is reasonably consistent.
  - Use style rules based mainly on **good sense** ... not just on precedent. Don't try to be consistent with past documents, unless there is a good reason to do so.

FAMILIARITY

- Use terms, abbreviations, and symbols that are *familiar to users*. Even if you are writing for highly technical people who are familiar with technical terms, they are still more familiar with everyday words.
  - *Say* "All the fish died."
    Not "The biota exhibited a one hundred percent mortality response.

- *Define terms* that may not be familiar to users. Define them the first time they are used, or in a Glossary. If you define them in the text, consider putting them in a box or otherwise calling attention to them, so that users can easily refer to them.

- When you use a new or complex word, make the other words in the sentence familiar. Don't use several new or complex words in the same sentence.

- Be wary about inventing or using unfamiliar codes, words, symbols, illustrations, etc., in situations where plain English will do. An article reported on a study by a police department.
  - They found that the 10-code used by many police departments, and popularized by TV and movies, causes more errors than does plain English. Using the code, 113 errors were made during 200 randomly selected police calls. Even after a refresher course, 85 errors were made in 200 calls. Using plain English, there were only 14 errors in 200 calls.
  - It would seem that the codes would at least save time. But the study found that using plain English instead of code saved nearly 3 hours each day. So now they say, "We've got this fellow in back of the van," instead of "10-95."

- The above study has a moral that goes beyond its specific topic. Don't introduce unfamiliar language of any sort, unless you have a really good reason to do so. If you must introduce unfamiliar language —
  - Link the unfamiliar with the familiar.
  - Provide a training course.
  - Even after the course, allow for ample learning time.
  - Introduce the new system in non-critical operations, so as to obtain experience with it — to see how it might be improved, or if it should be used at all.
  - Provide for both human and automatic error checking, if possible.
  - After a time, do a study to see if your system of codes is really working any better than the 10-code system discussed above.

INTEGRITY

* "The great enemy of clear language is insincerity."
  (George Orwell)

* Tell the truth, the whole truth, and nothing but the truth.

* Some unclear writing is intentional, to ignore, gloss over,
  or cover up unpleasant details, or to try to make users
  believe things that just aren't true.
  * A car spokesperson insisted that the cars being sold
    didn't break down. It was admitted, however, that
    sometimes they "failed to proceed."
  * Supermarkets are terming spoiled fruit and vegetables
    "distressed produce."
  * Instead of admitting that friends were on drugs, a
    student was quoted as saying that they had a
    "pharmaceutical preference."
  * And a psychiatric center came up with a novel way of
    admitting that a patient had been beaten to death. The
    patient, they said, "died of inappropriate physical
    abuse." Apparently appropriate physical abuse would
    have been okay.

* Don't ignore unpleasant details.

* Don't gloss over or cover up unpleasant details with
  vague statements, euphemisms, or generalities. One
  book advised that you should emphasize what things **are**
  or **will be**, instead of what they aren't or won't be. And
  that you should emphasize what you **can** and **will do**, not
  what you haven't done, can't do, or won't do. It certainly
  is important to stress the positive, or nothing would ever
  get done. But to ignore or hide the negative is just plain
  lying. Give a true version of both sides of the picture.

* Don't hide unpleasant details by devoting just a few lines
  to them, while devoting pages to the pleasant details.

* Don't hide unpleasant details by burying them in a mass
  of boring details that users are likely to skip over.

* Don't have "secret subjects," perhaps to hide who did
  something.
  * **Say** "After analyzing the data, **Pat Smith and I
    recommended** ...."
    Not "After analyzing the data, **the recommendation
    was made to** ...."

- Don't use the word "they," unless it's perfectly clear who
  they are.
  - **Say** "**Jack and Jill** went to the meeting. Then they
    went to lunch."
    Not "**They** went to the meeting. Then they went to
    lunch."
- Don't lie, exaggerate, or put on airs. Not even once! If
  you do, users may suspect everything else that you say.
  - "When one is caught bluffing too often, it is difficult not
    to wonder whether or not that individual is bluffing on
    the next issue." (Joseph M. Fox)
- If you are reporting on a study, don't just tell about the
  good results. Also tell about the blind alleys,
  inconclusive results, and outright failures. They may be
  just as important for users to know, as the good results.
  Moreover, if you don't mention obvious alternatives, users
  may wonder if you tried them ... and if not, why not.
- If your final results are inconclusive, say so. Don't make
  it sound as if everything has been neatly tied up.
- If a study or a project failed, admit it and tell —
  - Why it failed.
  - What, if anything, can be salvaged from it.
  - What, if anything, can, or is, or will be done.
- If you don't know something, **say so**. There is nothing
  wrong with not knowing. But there is something terribly
  wrong with not acknowledging, to others and perhaps
  even to yourself, that you don't know. If you don't
  acknowledge it to others, it's deceitful. If you don't
  acknowledge it to yourself, it's immature.
  - "It wasn't until quite late in life that I discovered how
    easy it is to say, I don't know."
    (W. Somerset Maugham)
  - "I was gratified to be able to answer promptly and I did.
    I said that I didn't know." (Mark Twain)
- Don't use many hedge words to enable you to take both
  sides of a position and offend no one.
  - **Say** "This **won't work**, because ...."
    Not "This **may perhaps** be a good idea **if** it is
    implemented correctly and **if** the timing is correct. But
    it seems that there **appear to be** some problems that
    may arise."

ORGANIZATION

- "The human mind can seemingly understand any amount of complexity ... as long as it is presented in small, simple chunks that are well organized."

- Organize the document into **short sections**, each with a single idea.

- Start each major section on a **new page**. This will cost some more in paper and printing costs. But it will help users to —
  - Understand the organization of the document better.
  - Find sections more easily and more quickly.
  - Reduce errors.
  - And it will make it simpler for you to revise the document.

- Combine related information in **one place**. Refer to it elsewhere, if it's also needed there.

- **Minimize cross-referencing**. When needed, give the page number of the referenced information, not just the section name or number.

- Use tables, charts, checklists, lists, pictures, etc., instead of, or in addition to, text.

- Organize the document as follows —

Discuss First	Before
General Interest Topics	Special Interest Topics
Principles	Details
Important Details	Minor Details
Permanent Issues	Temporary Issues
Matters That Affect Many People	Matters That Affect Few People
Etc.	Etc.

- *Put the most important information first*. There are two reasons why writers may not do this.
  - First, their pride may be hurt by the thought that users won't read every word of their document. But users are busy people. Your document is just one of many that they have to look at, not to mention all of the other things that they have to attend to. Do *you* read every word of every document that is sent to you?
  - Second, writers may want to end dramatically, and so may save the most important information or conclusions for last. But if you do this, it may not be read at all, or it may not be given proper attention by users.

POSITIVENESS

- **Positive sentences are clearer than are negative sentences**. According to one study, it takes about 50% longer to understand a negatively worded sentence (with words as "no," "not," "except," and "unless"), than it does a positively worded sentence. And the negative sentence may easily be misunderstood.
  - **Say** "**Consult** your supervisor before taking action."
    Not "**Don't** take action without first consulting your supervisor."
  - **Say** "**Positive** sentences **increase** readability."
    Not "**Negative** statements **do not increase** readability."
  - **Say** "**Leave** this box **blank** if you already have the book."
    Not "**Do not write** in this box if you already have the book."

- **Double negatives can be very hard to understand**.

Positive	Double Negative
only when	not ... until
only if	not ... except
only if	not ... unless
accept	not ... reject
agree	not ... disagree
legal	not ... illegal
succeed	not ... fail
present	not ... absent
like	not ... unlike
passed	not ... fail
usually	not ... often

  - **Say** "Do this **only when** needed."
    Not "Do **not** do this **until** it is needed."
  - **Say** "The contract becomes valid **only if** everyone approves."
    Not "The contract does **not** become valid **unless** approved by everyone."
  - **Say** "They were **present**.
    Not "They were **not absent**."
  - **Say** "X is **like** Y."
    Not "X is **not unlike** Y."
  - **Say** "They **passed** the test."
    Not "They **did not fail** the test."
  - **Say** "Pat is **usually** on time."
    Not "Pat is **not often** late."

- Here are some other positive words to be used in preference to negative words. The positive words are easier to understand.

Say	*Not*
minor	not important
dishonest	not honest
forgets	doesn't remember
ignores	doesn't pay attention to
suspects	doesn't trust in
dislikes	doesn't like

- If you do use "no," "not," etc., consider emphasizing the negative word with underlining or with bold italic type. Otherwise, users may overlook the negative word and thus may interpret the sentence in exactly the opposite way to what you intended.
  - "It is *not* ready."
  - "*Never* do that.

- People more easily understand sentences with the word "more" or "over" ... than with the word "less" or "under" or "at least."
  - *Say* "Push the switch only if the engine speed is *more than* 3000 rpm."
    Not "Push the switch only if the engine speed is *at least* 3000 rpm."
  - *Say* "Check this box if your trip is *over* 100 miles."
    Not "Check this box if your trip is *at least* 100 miles."
  - *Say* "Declare the interest from savings if it is *more than* 100 dollars."
    Not "You need not declare interest from savings if it is *under* 100 dollars."

- Phrase questions so that users are asked to check, underline, or circle what *does* apply rather than what does not apply.
  - *Say* "Check this box if you are *over* 21."
    Not "Check this box if you are *not under* 21."

- Avoid double questions.
  - *Say* "Are you over 21? Are you over 120 lbs. in weight?"
    Not "Are you over 21 and over 120 lbs. in weight?"

- Putting negative and positive phrases in opposition to one another, makes for good contrast.
  - "Not charity, but simple justice."
  - "Ask not what your country can do for you ... ask what you can do for your country." (John F. Kennedy)

PRECISENESS

- *Be direct*, so that you can't be misunderstood. It's more polite and also more fair, to others as well as to yourself, to say exactly what you mean.

- *Use specific language about measurements*. Don't say things like, "It takes a long time to do that job." One person may assume that "a long time" means an hour, while another may assume that it means a week. The same goes for costs, quantities, and all other measurements. If you can't give a precise figure, then give an approximate figure. Or give an approximate upper and lower bound, such as, "It usually takes from about 5 to 8 hours to do that job."
  - *Say* "This information would cost *1000 dollars* to obtain."
    Not "This information would cost *too much* to obtain."
Say	*Not*
Daily	Often
Turn until hand-tight	Turn a few turns
At least 3	Several
Between $7000 and $9000	Under $10,000
On July 23	In the near future
A 23% gain	A sizable gain
Before Thursday	As soon as possible

- Here is an example of how to be increasingly specific —
  - Someday
  - In the near future
  - Soon
  - As soon as you can
  - In a couple of weeks
  - Within 10 days
  - By February 2

- When presenting facts, data, etc., beware of the following shortcomings —
  - Superficial analysis of data.
  - Conclusions not justified by the evidence presented.
  - Failure to recognize and justify assumptions.
  - Failure to qualify tenuous assertions.

- *Use specific language about people*.
  - *Say* "Terry can *discuss problems* with Pat."
    Not "Terry can *relate* to Pat."
  - *Say* "Terry *refuses to speak* with Pat."
    Not "Terry *won't deal* with Pat."
  - *Say* "Terry *makes many typing errors*."
    Not "Terry *causes problems*."

- Anaphora are words or phrases that refer back to a previous word or words — "he," "she," "it," "they," "there," "that," "which," "the above," "defined earlier," "the second paragraph," etc. Such words slow down the user and often sow confusion. To avoid this problem, either repeat the word(s) that you are referring to, or reword the sentence. See the next item for an example.

- Often a source of confusion is *who* did (or is doing, or will do) *what*.
  - "Mary told Sue that she would manage the job." Who is to manage the job, Mary or Sue? To avoid confusion, repeat the name ... "Mary told Sue that she, Mary, would manage the job."

- Clearly identify —
  - *Specific Procedures* — So that they won't be confused with general discussions.
  - *Examples* — So that they won't be confused with procedures or data (and vice versa).
  - *Opinions, Assumptions, Recommendations, Value Judgments, And Estimates* — So that they won't be confused with facts (and vice versa). And tell how the opinions, etc. were arrived at.
  - *Proposals* — So that they won't be confused with actual plans (and vice versa).
  - *Preliminary Information* — So that users won't assume it is "final" and base their plans on it.

- Be careful about using loaded wor
  great [*bad*] equipment ...," "The *we*
  function ...," "The *crystal clear* [*inco*
  memorandum ...." Whether you use w
  of criticism, *back up the words with*
  Otherwise, you may mislead users.
  - They may read more, or less, into your
    criticism than you intended.
  - They may have different criteria for judg
    do. If you don't state the criteria that you
    to judge the person or thing, users have no
    knowing if your criteria are the same as the
    instance, equipment can be judged on reliab
    to purchase, cost to operate, cost to maintain
    energy efficiency, size, appearance, safety,
    durability, versatility, and other factors. Just sa
    that the equipment is good or bad, without going
    detail, may be worse than saying nothing about it.

- Clearly identify the *relationship* between topics in yo
  document.

- The *order* in which you discuss topics may influence
  people.
  - They may assume that the topics you discuss first,
    are the most important. Are they?
  - If your document is long, users may read just th
    few sections, and not even get to the last se
    Or they may just read the first and last se
  - Events and actions may be sequenced b
    as they really occur ... order of impor
    frequency of usage ... etc.   *ing and*

- To minimize the above problem —*opics.*
  - Explain at the start, the orde
    why.
  - Clearly identify the rela*

to each topic
a few
you devote
that the
you
t

71

ds, such as "That
/ [poorly] designed
mprehensible]
acts.

praise or
ng than you
are using
way of
rs. For
ity, cost

ying
into

r

- *Document data first*, and keep the data up-to-date. Document interface data used by many persons, separately from internal data used by just a few persons. The interface data are usually shorter, and are needed sooner, than are the internal data.
  - "Until late in the project, there wasn't nearly enough attention given to the documentation of the data as distinguished from the documentation of the programs. I participated in a couple of projects, where the real problem that underlay the difficulties we were having was that the global data base wasn't under any central control and central documentation. I think that given my choice of either one or the other, I would have documentation of the data rather than of the programs. Eventually we will have to produce both, but we had better get the documentation of the data out first." (From a debriefing report by the author of this handbook.)
  - "We produced a specification that defined both the internal workings and the external interfaces of a module in a single document. They should be in two separate documents. The interface information is usually smaller, and it's where the problems are. It should be separated from the internal information, so that it can be distributed very rapidly and kept up-to-date. It should not be combined with the longer and more detailed internal information on the program, which would slow up dissemination of essential interface information." (From a debriefing report by the author of this handbook.)

- Include a note asking for suggestions on how a periodically issued document could be improved. List possible ideas for changing the document, and ask users to check those ideas about which they feel strongly for or against, indicating their thoughts on the matter. For instance —
  - Change the content or the format.
  - Add new sections or delete existing sections.
  - Combine with other documents.
  - Issue at different times.
  - Eliminate the document.
  - List the various sections of the document, and ask users to check which sections they use Always, Sometimes, Rarely, or Never.

- Ask those users who no longer need or want the document to let you know, so that you can take their names off the distribution list. Include a self-addressed sheet or envelope for this purpose. Or say that unless each recipient tells you by a certain date that they do want to continue receiving the document, they will be taken off the list.

- Consider if you should reissue the entire document, rather than update pages that users are supposed to insert in their copy of the document. Reissuing the entire document may cost more, but it has many advantages —
  - Asking users to insert update pages involves time on their part ... probably costing far more money than is saved by not reprinting the entire document.
  - Users may insert update pages incorrectly.
  - Users may not bother to insert update pages at all.
  - Users can never be really sure that their document is up-to-date. Who is going to bother to check every page with the Page Inventory (assuming that there is a Page Inventory, and that it is current)?

- Have a Page Inventory listing each page in the document, and the issue dates of revised and new pages.

- Discuss changes and additions from the previous edition of the document in a Cover Letter, or in the Abstract or Introduction of the document.

- Send a sequentially numbered Cover Letter with each update. Ask users to file the Cover Letters at the beginning of the document.

- Have vertical lines in the margins of pages to indicate changed and new material, or next to the page number to indicate that most or all of the page is changed or new.

- Issue a new Contents with every set of update pages.

- If you need to criticize or complain about someone or something —
  - Find something **nice** to say also.
  - Offer suggestions as to what **should** be done.
  - Criticize the **action** or the **thing**, but **not** the person.
  - Make your criticism **specific** — to limit its impact on those involved, and also so that they will know exactly what you want.
    - **Say** "The equipment, which is currently on a 3 foot high table, should be set on a table 4½ feet high." **Not** "The incompetent person didn't set up the equipment correctly."

- Don't use sarcasm. Some people may take offense at it. And other people may not even realize that it's sarcasm. They may take it literally, thus interpreting your words as far worse than you intended.

- Humor and cartoons can be useful in some types of writing. But —
  - Be careful that they won't offend or confuse anyone.
  - Make sure that they won't downgrade the importance of your document, or of you as the author.

- Don't anger users. Be careful not to sound dogmatic or aggressive. You may be a highly respected authority. Your opinions may be the best in the world. *But others are entitled to their opinions*.

- Show some *feeling* about what you are writing, especially if the purpose of your document is to evaluate something or to sell it. If you think that it's great (or terrible), say so with feeling. A few heartfelt words will get your message across far better than would loads of dry words.
  - *Say* "This is a truly innovative idea that I am sure will be of great value."
    Not "This is a good idea that should prove useful.

- Be polite with positive, encouraging words. Minimize negative, critical, discouraging words. However, don't gloss over things. Be as clear as you can be about negative aspects.
  - *Say* "You *did not* ...."
    Not "You *neglected* to ...." or "You *failed* to ...." or "You *forgot* to ...."
  - *Say* "You *say* that ...." or "You *indicate* that ...."
    Not "You *claim* that ...." or "You *assert* that ...." or "You *allege* that ...."

- Don't insult or try to intimidate people. Words as the following are better left out.
  - "It should be obvious that ...."
  - "It is clear that ...."
  - "Naturally ...."

- Especially bad are words starting with **un** —
  unsuccessful ... **un**happy ... etc. Better to use **not**.
  - "I'm **not** interested" is more emphatic than "I'm
    uninterested."
  - Or, express a negative in positive form — "trifling"
    instead of "not important" ... "ignores" instead of "does
    not pay attention" ... "dislikes" instead of "does not
    like," etc.

- Consider if you want to emphasize the **positive**, or the
  negative —
  - "The tank is **half full**."
    is the same as
    "The tank is **half empty**."
  - "**Thirty-eight** of the forty tests **succeeded**."
    is the same as
    "**Two** of the forty tests **failed**."

- Enthusiasm is nice. But too much can be ridiculous.
  - **Say** "I am happy to offer you a position on our staff of
    educated and skilled persons. The work is challenging
    and pays a good salary."
    Not "I am **extremely** happy to offer you a position on
    our staff of **highly** educated and **exceptionally** skilled
    persons. The work is **extraordinarily** challenging and
    pays a **very** good salary."

FOOTNOTES

- Put notes as footnotes on the bottoms of the pages to which they apply, in smaller print than the text. Thus users can easily glance at the notes and see if they are interested in reading them. If you put notes at the end of chapters or of the document, it takes users time and effort to find them, and breaks up their train of thought. Most users probably won't bother to look for them, but they may have a lingering doubt that they should have looked at the notes, and perhaps they should have. This may leave users with a bad aftertaste for your document (and rightly so).

WRITE-IN SPACE

- Leave space for users to write in specific information needed while using the document — for instance, their computer job number. The following was on the Cover Page of a foldout card —

- Name of Application
- Administrator
  - Name
  - Phone
  - Room
- How to Access
- How to Request Paper Forms
- Where to Send Paper Forms and Attachments

FORMAT

Note — The format on this page was used for a computer document, so it might not be applicable to other subjects. However, it's the **idea** ... not the details ... that is being illustrated here — to have a uniform format where applicable.

• Use a standard format to specify information for similar items. Put section titles in the left-hand margin. For instance —

```
PPTP          ⎧ PPTP  ⎫
PPTR          ⎪ PPTR  ⎪
PRINT     $ ⎨ PRINT ⎬  filecode, logical-unit-designator
PUNCH         ⎪ PUNCH ⎪
READ          ⎪ READ  ⎪
TYPE          ⎩ TYPE  ⎭
```

Examples   • $ PPTP A4,B2S
            • $ PPTR 40,A2R

Purpose   • To associate a filecode with a peripheral device
              (normally used with BMC jobs):
                PPTP - Paper tape punch.
                PPTR - Paper tape reader.
                PRINT - Output to printer if not via SYSOUT.

• Within the format, progress from general information for all users, to detailed information for specialized users.

• Put the formats for computer commands (or whatever), directly in the headings ... not buried in the text. For instance —

byte (constant, *constant*, ...);   8-bit
word (constant, *constant*, ...);   16-bit

• **Underlined items** are optional.

• **byte** and **word** — Store data items of the size indicated, at the point of occurrence in **.text** or **.data** sections. Must **not** be used in **.bss** section.

PAGE NUMBERING

- Number all pages, even in early drafts. Do this for your own benefit as well as that of people reviewing the drafts, so that the pages don't get mixed up or lost.

- Probably the simplest and most exact way to number pages is to start with the very first page (the Outside Front Cover) as 1, and then to number following pages 2, 3, 4 .... Complex methods showing volume, issue, chapter, and finally page number may be of interest to authors and editors, but such methods just confuse and slow down users. It's easier to find page 186 than page III-15-5-2.3!

- Put on the front page of a stapled document, "Page 1 of 20 pages," so that users will have an easy way of knowing if all of the pages are there.

- If at a later time you need to add pages and you would rather not change existing page numbers, you can number the added pages as 10a, 10b, ...

- If the document is typeset, use large size, bold italic type for page numbers to make them stand out better.

- If you put page numbers in the center (top or bottom — preferably the top) of the page, then you won't have to be concerned about whether the page will be a right- or left-hand page when the document is printed two-sided — as you would have to be concerned if page numbers were typed on the side of the page. However, page numbers on the outside, top corners of pages are easier for users to find. A word processing system may be able to automatically place page numbers on the outside top corners correctly on the right- or left-hand sides of the pages for two-sided printing.

TITLE/ACRONYM

- The title is the first thing that people see. It may be the *only* thing, as in an index of titles. So pay attention to it. The best document is of little value if few people are motivated to get it and read it.

- Make the title *clear and meaningful*. Not clever or cryptic. People want to know — What can this do for me? Tell them.

- Don't be afraid of long titles. The main thing is to tell people what they should know.

- If your title is more than two or three words, consider using a one-word identifier, or an acronym, to identify the document.
  - *Say* "QUEST: QUALITY ENVIRONMENT FOR SYSTEM TRACKING"

- Tell if this is a draft, proposal, order, specification, handbook, etc.

- Identify the overall system, and the specific subsystem.

- Make the title *specific*.
  - *Say* "A COMPREHENSIVE OVERVIEW OF *60* PRODUCTIVITY TECHNIQUES"
  Not "A COMPREHENSIVE OVERVIEW OF PRODUCTIVITY TECHNIQUES"

- Use titles in the form of *questions* that the user might ask.
  - *Say* "CAN I GET A COLLEGE LOAN?"
  Not "STUDENT ELIGIBILITY"
  - *Say* "WHAT OTHER FORMS OF AID CAN I GET?"
  Not "OTHER FORMS OF AID"

- Focus on the **results** to the user. Not on the work that you did. You can help do this, by including descriptive **keywords** in the title. Put the keywords **first** in the title.
  - **Say** "PROJECT XYZ: A PRELIMINARY REPORT"
    Not "A PRELIMINARY REPORT ON PROJECT XYZ"
  - **Say** "THEORETICAL PHYSICS: AN INTRODUCTION"
    Not "AN INTRODUCTION TO THEORETICAL PHYSICS"
  - **Say** "COFFEE AND HEARTBURN: A POSSIBLE LINK"
    Not "A STUDY OF THE EFFECTS OF COFFEE ON THE DEVELOPMENT OF HEARTBURN"
  - **Say** "THE SUMMIT PLANNING BOARD: ITS PURPOSE AND ORGANIZATION"
    Not "THE PURPOSE AND ORGANIZATION OF THE SUMMIT PLANNING BOARD"
  - **Say** "ELECTRIC MOTORS: HOW THEY WORK AND SOME EXPERIENCE WITH THEM"
    Not "HOW ELECTRIC MOTORS WORK AND SOME EXPERIENCE WITH THEM"

- Make the title **interesting**.
  - **Say** "IS STRESS KILLING YOU?"
    or "STRESS: IS IT KILLING YOU?"
    Not "THE EFFECTS OF STRESS"

- Use **subtitles**.
  - **Say** "SALT: WHAT IT DOES IN YOUR BODY
    Salt Helps To Keep The Blood Neutral, Distributes Water, And Enables The Muscles To Function Properly. But Too Much Salt May Cause High Blood Pressure in Susceptible People, As Well As Other Problems"

DESIGN — BASIC SECTIONS

COVER — OUTSIDE FRONT

Many documents have a beautiful photograph or a fancy
design on the outside front cover, but no *information*
except for the title (which itself may not be very
informative). Such covers do not help users. Put on the
outside front cover, information such as —

- Title
- Company Name / Logo
- Date
- Issue / Version Number
- Name / Number of Department
- Address / Phone
- Hotline Phone To Call For Assistance
- Names / Editions of related documents
- Other documents that this document supersedes
- Abstract
- Contents
- Copyright / Restrictive Notice
- Page Number of Contents and Index
- If readers need to take some action, call their attention to
  this in the Abstract or right under the Title. Write in
  big / bold letters — "Action Required By (Date)."
- Have a "family resemblance" on the front cover, between
  documents on the same subject — same logo ... similar
  titles ... etc.

One reviewer of this book felt that too utilitarian a cover
might give an impression of lower quality. I disagree. I feel
that the "form should follow the function" ... that the function
of the outside front cover is to *inform* users of what's
inside and where to find it. However, if you agree with that
reviewer, you could put information such as the above on
the inside front cover, or the outside back cover.

COVER — OTHERS

- Put useful information on the inside front cover and the
  inside / outside back covers, such as an abbreviated
  Index, or a summary of essential information.

CONTENTS

- Have a **CONTENTS**. Put in page numbers (some people don't, for reasons known only to them).

- Put the **CONTENTS** on the outside front cover, directly under the Title and identifying information.

- For a loose-leaf book, consider having a plastic, see-through pocket on the outside front cover for a replaceable Title/Contents page. Thus you can issue a new Title/Contents page when you reissue the document contained in the loose-leaf book.

- Have lists of figures, tables, illustrations, etc.

- Avoid conflicting number systems. For instance —
    1. Table 3
    2. Table 4
  In the above example it would be better to say —
    1. Table 1
    2. Table 2
  Or just list the tables —
    Table 1
    Table 2

INDEX

- Have an Alphabetic (and perhaps a Functional) **INDEX**, in addition to the **CONTENTS**.

- List the same topics in two or more places in the **INDEX**. For instance, you might list "Stereo Radio" in two places ... in one place as "Stereo Radio," and in the other as "Radio, Stereo."

- If you have several page references to a given topic, indicate which is the **main** page reference by putting that page number in bold italic type, or underlining it, etc. For instance, "Stereo Radio ... 3, **10**, 19, 50."

- The above point implies that you do, in fact, discuss every important point **fully** in one place. Don't leave yourself open to the following criticism — "When I try to find something through a manual's index, too often my search ends in total frustration."

- Have an abbreviated **QUICK INDEX** on the outside back cover.

ABSTRACT

- Very briefly state the objective of the document ... some background information ... the problem or goal that is discussed ... the scope and approach ... the results ... and the conclusions. The user should be able to read just the **ABSTRACT** and not miss any key point.
Note — What has been discussed is an *informative* type abstract. Such an abstract isn't possible for long and involved documents. For such documents, write an *indicative* type abstract that discusses the type of information in the document.

- Tell *who* should use the document, and *why, when, where,* and *how*.

- Have a "selling" message, telling why people should read the document. Emphasize the *benefits* of reading it, not just its features. Will it help them to do their job better, faster, more accurately, more easily, more successfully?
Example — A stereowide control on a portable radio is the *feature*. Realistic stereo separation is the *benefit*.

TO THE USER

- *Note* — The **ABSTRACT** is usually shorter, and should be on the outside front cover. The **TO THE USER** section should amplify on the **ABSTRACT**, and should be on the inside front cover.

- Explain the purpose of the document. Even if it's obvious to *you*, it may not be obvious to *others*. Or they may misunderstand your purpose. For instance, what you intend as an *order*, they may take as a *recommendation*. Or what you intend as a discussion of your *opinions*, they may take as *facts*.

- Tell if there are other documents that users should read or have read ... and if this document supersedes any documents.

- Should the user — Read the document consecutively? Read whatever chapter he or she needs at the time? Read the introductory chapter(s), and then whatever chapters are needed? Etc.

- If your document provides training on a subject, suggest to users how to approach it. Should they learn it on their own? Or attend a course? Tell how much time is likely to be needed.

- List phone numbers and addresses for more information.

- List acknowledgements to people who have contributed to the document.

OVERVIEW

- Overviews to each chapter or section may be useful, if the chapters or sections are long. However, if they are short, and/or if the headings and subheadings are clear, then an overview may not be needed.

INTRODUCTION

- An **INTRODUCTION** to a document or to a section of a document may present background material (what the work is about, reasons for doing the work, possible benefits, past work, etc.) or other material that will help orient the user to the document or section. It's different from the **OVERVIEW**, which summarizes the document or section. (But sometimes writers do include overview material in the **INTRODUCTION**.)

- *Note* — The distinctions between an **OVERVIEW**, an **INTRODUCTION**, and a **SUMMARY/CONCLUSIONS** (see next page) aren't clearcut. You don't need to have all of these sections, and you can include in them what you feel is best.

GLOSSARY

- Have an alphabetic **GLOSSARY** if needed.

SUMMARY/CONCLUSIONS

- "Effective immediately, I want every report specifically, directly, and bluntly to state at the beginning a summary of the unshakeable facts. ... The highest art of professional management requires the ability to smell a real fact from all others — and moreover, to have the temerity, intellectual curiosity, guts and/or plain impoliteness if necessary to be sure what you have is indeed what we call an unshakeable fact."
  (Harold S. Geneen, "The View From Inside," *Business Week*, November 3, 1973)

- *State conclusions specifically.* Don't bury them in the middle of, or at the end of your document. Have the *SUMMARY/CONCLUSIONS* right at the *start* of your document.

- Include —
  - Specifications, Bounds, Limits, And Assumptions
  - Review Of Major Facts And Ideas
  - Interpretation Of The Facts And Ideas
  - Recommendations For Action — *What should the user DO right now?*
  - Cost Trade-Offs
  - Lessons Learned
  - Pitfalls To Avoid
  - Evaluations
  - General Principles
  - Relationships

- Support conclusions with —
  - Data, Charts, Pictures
  - Examples
  - Experience
  - Testimonials by Experts and Users
  - Quotations by Experts and Users

- Have a chart or table that summarizes findings, recommendations, etc.

- Summarize many pages of data into a few small graphs that can fit on just one or two pages.
  - For instance, a 100-page monthly budget report was summarized on 20 small graphs, including — Balance Sheet ... Monthly Sales ... Monthly Pretax Profits ... Division Contribution to Pretax Profits ... Variable Income ... Return on Capital. This 1-page summary (with the 20 small graphs) was sufficient for most users. Those who needed more information could, of course, look at the details. And this summary gave an invaluable "Big Picture" to all users, which they could not get from the 100-page report.

- ***Note*** — If your document is long, you might want to have separate ***SUMMARY*** and ***CONCLUSIONS*** sections.

APPENDIXES

- Include in Appendixes, material that few users will need or want. Or if such material will take up many pages, include it in a separate, supplementary document. Refer to this supplement in your main document, so that users can request it if they want. Send copies of the supplement to those users whom you know will want it.

REFERENCES

- List documents and other references that you used in gathering information for your document. Also list others that you didn't use, but that may be of interest to users. Be nice to users and include a couple of sentences or so about each item, telling them what it's about and how it may be of use to them.

DESIGN — BASIC AIDS

EXAMPLES

- **Use examples liberally**. Use enough examples to cover the range of variations likely to be encountered by most users.

- Develop examples with as much care as the text ... perhaps with more care.

- **Test examples** to make sure that they are correct. All too often, examples and text contradict one another.

- Make examples **real-life situations** that users are most likely to encounter. And make them accurate, complete, and consistent with the text. Most users prefer to find an example that is similar to what they need to do, and then to follow it exactly or with whatever changes are needed. A good example or picture is worth a thousand words. But a poor example can nullify the value of many words of accurate text.

- Make examples as **complete** as possible ... not just unrealistic, partial sequences.

- Use both **specific** examples, and **overall** examples of **entire task sequences**.

- Work out some examples **in full**, so that users can see how solutions were obtained.

- If you are showing examples of filled out forms, write in by hand the parts to be filled in. When the document is printed, print these parts in red.

- If you are showing human/computer dialogue, use different typefaces to distinguish between the human and the computer. Make the typefaces as realistic as possible. Use an actual printout if practical. Preface each part of the dialogue with "You" (or "User") and "Computer." Put screen displays in boxes with rounded corners, that resemble screens.

- Place examples so that they are visible while the user is reading the related text — on the same or facing pages.

- Have a chart showing all of the possible components in a system, and their relationship to one another.

- ***Annotate examples*** to emphasize and explain key points, terms and abbreviations. Use arrows, etc. to point or otherwise relate to the appropriate areas of the examples.

ILLUSTRATIONS

- **Use photographs**. According to one study, photographs are remembered better than are drawings. However, drawings may show details better than photographs.
  - Use both **overall** photographs (or drawings) showing an entire scene or piece of equipment ... and **specific** ones showing details.
  - Use labels with arrows to point out specific features.
  - Use color.
  - If you use many photographs, you might want to read **Editing By Design: Word and Picture Communication for Editors and Designers** by Jan V. White (New York: R.R. Bowker Company, 1974). There are also other good books on the subject.

- Present only relevant information. Relevant information is the item of interest with sufficient surrounding background to permit that item to be located in the real world.

- Use "callouts" (explanatory words, phrases, or sentences) to explain what the user looks at in an illustration.
  - Use words and arrows to reference a small number of items.
  - Use numbers and lines to reference a large number of items. Arrange the numbers clockwise starting about 2 o'clock. Place corresponding numbers in the text.
  - Keep callouts outside of the body of a photograph, so as not to interfere with the realistic appearance of the photograph. Callouts are okay in the body of an illustration, since it isn't totally realistic to begin with.

- Show a detailed item enlarged in relation to the total illustration, when the item is a small portion of the total illustration.

- Orient the illustration the same way as the equipment.

LISTS

- Use **list form** whenever possible. Don't write out in sentence form, information that could be put in list form.

- Give all items in a list the same sentence or non-sentence structure.

- Make a long list of items more readable by skipping a line, or drawing a line, after every 3rd or 5th item or so.

- Put into columns, data to be compared or used together. Keep the columns **close together**, with 3 dots to relate corresponding items (for instance, names in one column, phone numbers in the other). Often columns are far apart; thus users must correlate the items, perhaps using a ruler to line up related items. This takes up the time of users, and they may easily make mistakes.
    - **Say** "A. Adams ... ext. 1111
        J. Jones ... ext. 2222
        S. Smith ... ext. 3333"
      Not "A. Adams   ext. 1111
        J. Jones   ext. 2222
        S. Smith   ext. 3333"

- List data in different ways for different uses.
  - Alphabetic Sequence
  - Numeric Sequence
  - Space (Layout of Physical Objects) — Right to Left, Left to Right, Top to Bottom, Bottom to Top, Inside to Outside, Outside to Inside, Or Not In Any Special Order
  - Time (Chronological Sequence) — Past to Present, Present to Past, Or Not In Any Special Order
  - Logical Sequence — Input to Output, Output to Input, Most Important or Most Numerous, First or Last, Or Not In Any Special Order.
  - Enumeration — Order of Importance, Occurrence, Familiarity, Etc.
  - Classification — By Importance, Weight, Height, Function, Cost, Size, Power, Etc.
  - Sequence of Use
  - Frequency of Use
  - Habitual Sequence
  - Order of Importance
  - Function
  - Familiarity
  - Complexity
  - Acceptability
  - Cause to Effect (or Effect to Cause)
  - General to Specific (or Specific to General)

- If an item in a list differs markedly from the other items, call attention to the differing item by one of the highlighting techniques. (But first consider why the differing item is in the list anyway. Perhaps it should be placed by itself.) For instance —
  Oranges
  Pears
  Pencils
  Grapefruits
  Peaches
  Plums

PERFORMANCE AIDS

- Performance aids are documents used for on-the-job instruction or reference — small looseleaf books, single or multipage cards, charts, booklets, etc. The moment that one sees people with "cheat sheets" (notes that they have compiled for on-the-job use) on their desk, taped onto equipment, written into the margin of a document, etc., the designer can identify a point where a performance aid may be useful.

- In addition to being a document in itself, a performance aid can also be part of a larger document. One or more pages of that document can be organized for quick, on-the-job use by users.

- Because performance aids are usually of compact and convenient size, they are more likely to be kept on the working area than are regular documents. Because they are clear, concise, and highly organized, they are easy to use. Thus they —
  - Aid memory.
  - Eliminate guesswork.
  - Reduce errors.
  - Save reference time.
  - Improve efficiency and speed ... and accuracy and reliability of performance.
  - Increase safety.
  - Reduce training requirements, and time.
  - Increase job satisfaction by making the task clearer, simpler, easier, and more successful.

- Performance aids are valuable for almost any task, but are especially needed if —
  - Task is critical.
  - Task is too lengthy, complex, or infrequently done to be remembered.
  - Other, more complex documents are less effective for on-the-job use.
  - Task is discretionary; for instance, a system that people will use only if they want to. A good performance aid will encourage its use.

- What kinds of information can a performance aid provide?
  - Directive — Step-by-step instructions.
  - Supportive — Calculation (e.g., rate or conversion tables), lists, interpretation (e.g., code translations), decision, or discrimination.

- Consider developing a performance aid for almost any purpose. Don't feel that a long, complex document will have more impact. On the contrary, the document may be skimmed over or even thrown away, while a performance aid may be kept on or near the working area and thus *used*.

- Explore possible physical forms —
  - Single/multipage foldout card
  - Leaflet/Booklet
  - Binder — regular/small size ... permanent (sewn/glued/stapled)/looseleaf/plastic spiral bound/wire bound ... top/side bound
  - Looseleaf rings without binder
  - Chart/Poster — pocket/desk/looseleaf/wall size
  - See-Through Overlay
  - Also consider — tabs ... plastic lamination ... outside see-through pocket on looseleaf binder for sheet with title/contents/index ... inside pocket for notes.

PROCEDURES

- *Note* — Also see the section on **WARNINGS** on page 113.

- State the *purpose* of the procedure. Don't assume that people know it.

- Tell *who* has to follow it, and *when*. Then tell *what* they should do. *Be specific*.

- Tell people *at the start* what they need to complete the procedure —
  - *Time* — So that they don't start a 60-minute procedure when they have only 10 minutes available.
  - *Skills, Experience, Etc*. — So that they don't undertake something that they can't handle.
  - *Information, Tools, Materials* — So that they aren't caught short in the middle.

- Present *first* the most important, and the most common aspects of the procedure.

- Tell if there are *exceptions* to part(s) or all of the procedure, the conditions under which the exceptions apply, and if any approvals are needed.

- Distinguish between *recommended and required* procedures. Write so that the choice of action is clear.
  - "You *should* do this within two weeks." Is this a *requirement* ... or just a *goal*?
  - "You *will* do this within two weeks." Is this an *order* ... or just a *description* of how long it's likely to take?
  - "You *must* do this within two weeks." There doesn't seem to be any doubt here.

- Have just *one instruction per sentence*.
  - *Say* "Stop the program. Then load the data."
    Not "Stop the program and then load the data."
  - Better still, use numbered (or bulleted) list form —
    "1. Stop the program
    2. Load the data."

- List the steps of a procedure in the *exact order* that they are to be performed.

- Fit all the steps of a procedure on *one page*, or on *two facing pages*.

- If the steps in a list must be carried out in sequence, say so and label each step with a number — 1. ... 2. ... 3. .... Also number them if individual items will need to be referred to from elsewhere. But if items are unrelated or don't need to be referred to, use bullets. When you use numbers or bullets, always list the steps **one under the other**. **Note** — I believe that numbers give an air of complexity. Avoid them when possible.

- Place **critical** information where it belongs **in** the list. **But** ... call attention to it at the **start** of the list. Otherwise, users may not realize that it's critical ... or they may not even get to it. For instance, if something is fragile, say so **before** you tell the user to move it.
  - **Say** "WARNING —
                THIS EQUIPMENT IS FRAGILE.
                PICK IT UP SLOWLY.
                SET IT DOWN GENTLY.
                1. Move the projector from room A
                   to room B.
                2. Put it on the table.
  - **Not** "1. Move the projector from
                   room A to room B.
                2. Put it on the table.
                WARNING —
                THIS EQUIPMENT IS FRAGILE.
                PICK IT UP SLOWLY.
                SET IT DOWN GENTLY.

- Use **directive** (active tense) statements.
  - **Say** "**Start** the machine ...."
    Not "The machine **should be started**."

- Give **feedback** on the effects of actions. That is, tell the user what **should happen** after he or she does something ... and what to do if nothing happens or if something else happens.
  - **Say** "Turn the lever to position C. The dial should then read 50. If it doesn't, do this ...."

- Point out **common errors** to avoid.
  - **Say** "Be careful that the lever is exactly on the C notch. Ensure that it is not between notches just before or after the C notch."

- Help motivate users to follow the procedure intelligently, by clearly stating the *reason* for it, and the *consequences* if it isn't followed.
  - *Say* "The reason for this procedure is to cool the unit *slowly*. If this procedure isn't followed exactly, the unit may be *damaged*, or may *explode*."

- Don't require users to follow step-by-step instructions from the beginning to the end of a procedure. Organize and chunk instructions so that users can easily bypass sections that aren't needed in certain situations, or that they already know.

- Avoid these mistakes —
  - Required action isn't clearly indicated.
  - Too many separate actions in a single statement (should be only one).
  - Too many thoughts, descriptions, or conditions in a single statement (should be only one or two).
  - Irrelevant or excessive explanatory material. Talks too much *about* the activity ... rather than about *what to do*.

TABLES

- *Note* — A discussion of tables, graphs, pictures, etc. could cover an entire book. Indeed, there are books on this subject. Get one or more books on it, if your writing requires many tables, etc. This section has some very basic information on the subject.

- Put tables, etc. near the text to which they apply, not several pages away.

- Explain tables, etc. in the text. Users may not bother to look at the tables, let alone to figure out what they mean. But on the other hand, some users may look only at the tables, etc., and not at the text. So each should, as far as possible, stand on its own.

- Clearly label tables in a uniform place (above or below the table). Use a different typeface, or underlining, to distinguish labels from standard text.

- Use a different typeface and/or size to distinguish the table from the text. Or box the table.

- Keep tables free of unnecessary clutter. Leave space between numbers, lines, etc. Don't crowd the table. If a table is complex, break it into two or more simpler tables.

- Don't break tables between pages. If you must do this, ensure that all parts are visible at the same time; for instance, that they aren't on reverse sides of the same sheet. Some possibilities —
  - Put the table on facing pages.
  - Reduce the table in size so that it can fit on a single page. But don't reduce it so much that it's hard to read.
  - Put the table on a foldout page. But note that foldout pages are expensive, inconvenient, and tear easily.
  - Break the table into *logical subsets*, one or more of which can fit on a single page or on facing pages. This may help to make the table more understandable and perhaps more usable, as well as more readable.
  Note — You can also put the entire table on one page, in much reduced form. Have boxes around the subsets, and tell which page it's on. Thus users can see how the subsets fit together.

- Arrange columns from left to right.

- Put the heading on both the top and the bottom, or on both sides.

- Use vertical lines or spaces to separate columns.

- Keep space between columns to a minimum, so that items in one column can be clearly associated with related items in the other columns.

- Use horizontal lines or spaces to divide the table into sections, perhaps at every 5th line. (Use heavier, thinner, or dashed lines, as appropriate, to chunk information.)

- Explain symbols in a Legend.

- Use columns for more important comparisons. It's easier to compare columns than rows.

- Boxes in a flowchart should usually go from left to right, as does text, not from top to bottom. The usual procedure is to show inputs on the left and outputs on the right. **Note** — Flowcharts may seem useful. But studies indicate that many users find them confusing (perhaps because they aren't used to them), and don't use them. Several studies have indicated that good verbal instructions result in better performance than is obtained with flowcharts. Moreover, flowcharts are costly to create and to revise, which means that they may quickly become out of date. Try to use some other method — text, lists, decision tables or trees (if you want to consider using these, find a book that describes them), etc. Make sure that your method is **simple to use**, and **simple to revise**.

- Identify units of measurement, and any units that vary from the other units used. For instance, if ten items in a list are in feet and one or two are in meters, call attention to the items that are in meters.

- Put units of measurement in the column headings, rather than in the data fields.

- Round to whole numbers. Don't use too many decimal places.

- Emphasize similarities and differences.

WARNINGS

- *Note* — Get *medical*, *legal*, and *editorial* consultation on how to write and present warnings.

- Everything in this book applies to warnings. Especially, see *PROCEDURES* (page 108), and *COLOR* (page 120).

- Because warnings are so important, you should devote ample time and attention to —
  - Thinking of *every imaginable warning*.
  - Writing it so that it will be clearly understood.
  - Highlighting and placing it so that it can't be missed.

- Put warnings —
  - On the equipment itself.
  - In the manual for the equipment.

- Don't assume that *anything* is so obvious that you need not warn people about it. It may *not* be obvious to them. And even if it is, they may be preoccupied, tired, etc. Thus the warning will serve as a useful reminder to them.

- *Always test a warning on several users to whom it is aimed, in their actual working environment ... no matter how clear, simple, and obvious it appears to you*.
  - An example of a notice that appeared "obvious" but wasn't, was the sign posted next to an elevator — "PLEASE WALK UP ONE FLOOR, WALK DOWN TWO FLOORS FOR IMPROVED ELEVATOR SERVICE." It has no difficult words, yet people were found walking up and down the stairs looking for a floor from which they could get improved service, only to find the same sign on each floor. An alternative wording is clearer — "TO GO UP JUST ONE FLOOR OR DOWN JUST TWO FLOORS, PLEASE WALK."

- *WARNINGS SHOULD COME FIRST.* Users may skim over, or even ignore the text, and thus might miss the warning if it's buried within, or follows the text.
  - *Say* "DANGER —
        VALVE MAY BE HOT
        1. Open valve A.
        2. Drain the water."
    Not "1. Open valve A.
        2. Drain the water.
        DANGER —
        VALVE MAY BE HOT."

- Call attention to a warning —
  - See *HIGHLIGHTING* on page 126.
  - Use a heading. Perhaps skip a space between each letter and also underline it once or twice.
    For instance —

ATTENTION

A T T E N T I O N

<u>A T T E N T I O N</u>

- Place asterisks, or some other symbol, all around the warning. For instance —

* *

<u>A T T E N T I O N</u>

text of warning

* *

- Use shaded areas.

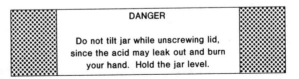

DANGER

Do not tilt jar while unscrewing lid, since the acid may leak out and burn your hand. Hold the jar level.

- Be consistent in how you label warnings.
  - Inform users, at the start of your document, of the warning system that you are using. For instance —
    DANGER — Possibility of personal injury.
    CAUTION — Possibility of service interruption.
    WARNING — Possibility of equipment damage.
  - If two or more possibilities exist together, label the notice with the *highest* priority.

- Warnings may be of several types. They may tell the user to —
  - *Prepare* themselves before doing something ... as when putting on safety glasses.
  - *Observe* something *usual* ... as a meter reading.
  - *Watch* for something *unusual* ... as a pipe leaking.
  - *Do* something *usual* ... as tightening some screws.
  - *Do* something *unusual* ... as cleaning up if a pipe has leaked.
  - *Not* do something *usual* ... as not touching a hot pipe because it's always hot.
  - *Not* do something *unusual* ... as not touching a pipe if it's hot. (This is also a *do* action, since the person would first have to gently touch, or otherwise test the pipe to determine if it's hot.)
  - Combination of two or more of the above.

- Make the warning *direct and personal*. It may, and usually should be accompanied by a more detailed explanation. The instructions on one piece of equipment stated —
    > *DANGER: The batteries in the AN/MSQ-55 could be a lethal source of electrical power under certain conditions*.
  On the equipment itself, someone had written in large red letters —
    > *LOOK OUT! THIS CAN KILL YOU!*

- But neither of the above warnings told *under what conditions* the equipment could be dangerous. This is like the shepherd who cried *Wolf!* so many times when there was no wolf, that no one paid attention when there really was a wolf threatening the sheep. A warning should state specifically the conditions that might cause danger. For instance —
    > *DANGER!*
    > *LOOK OUT!*
    > *THIS CAN KILL YOU IF YOU*
    > *TOUCH THE BATTERY TERMINALS*

- Negative words ("no," "not," "don't," etc.) lower readability of regular sentences. However, they can be useful in warnings. For instance —
    > *DON'T TOUCH THIS!*
    > *IT'S HOT!*

- Break lines logically.
  - *Say* "Heavy equipment.
      Can cause severe injury."
    Not "Heavy equipment. Can
      cause severe injury."

- As appropriate, do *not* soften words to avoid alarming people.
  - *Say* "Dangerous gas.
      Will damage eyes if released."
    Not "Dangerous gas.
      May irritate eyes if released."
  - *Say* "High voltage.
      Will cause severe injury or death."
    Not "High voltage.
      May cause severe injury or death."

- Don't use contractions like "it's," "don't," "can't," or "won't" in warnings. "It is," "do not," "cannot," and "will not" are more emphatic. (But I feel that "it's," "don't," "can't," "won't," etc. are more informal and thus are preferable, in regular sentences.)

- Be precise in your use of *verbs*.
  - Don't say "remove" if you mean "disconnect."
  - Consider if "pull" or if "jerk" is more descriptive.

- Be precise in your use of *adjectives*.
  - *Say* "The *red* handle ...."
    Not "The *dark* handle ...."

- Be precise in your use of *nouns*. For instance, consider if "handle" or "lever" is more descriptive ... "fastener" or "staple" ... etc.

- Underline, or put in bold italics, and perhaps also put in full caps, words like "no," "not," etc., to emphasize them. Otherwise, users may overlook the negative word and may thus interpret the sentence in exactly the opposite way to what you intended. This is especially likely to happen in emergency circumstances when people are rushed and tense.
  - *Say* "Do *NOT* touch this."
    Not "Do not touch this."

- The Westinghouse Electric Corporation has developed a handbook with a step-by-step approach to producing product safety labeling, *Product Safety Label Handbook*, MB3699, by J. F. Gormley, et al. It can be purchased from —

  > Westinghouse Electric Corporation
  > Customer Service, Printing Division
  > Forbes Road
  > Trafford, PA 15085
  > Phone: 1-412-829-6318

- Some highlights of the labeling —
  - *Example* —

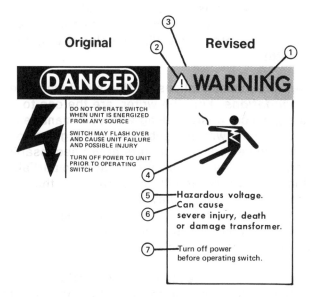

1. *Signal Word — DANGER, WARNING, CAUTION,* or *NOTICE* appears at the top.
2. *Hazard Alert Signal* — ⚠ appears with signal words *DANGER, WARNING,* or *CAUTION*. It's an international standard that means *Look Out!*
3. *Color — DANGER* (Red), *WARNING* (Orange), *CAUTION* (Yellow), *NOTICE* (Blue).
4. *Symbols and Pictographs* — These reinforce the verbal message. They also provide non-verbal information for illiterate or non-English-speaking readers.
5. *Identification of Hazard* — In bold type.
6. *Result of Ignoring Warning*.
7. *Avoiding Hazard.* Tells how to avoid injury.

COLOR

- Get the advice of an expert on visibility, contrast, readability, and how to avoid misunderstanding by color-blind people.

- A PMS book shows ink colors available, numbered according to the standard Pantone Matching System (PMS) used by printers. There is also a PMS book for paper colors. Buy them at an Art Supply store or from Pantone, Inc., 55 Knickerbocker Road, Moonachie, NJ 07074. Phone 1-201-935-5500. (An editor, or the Art or Drafting departments in your company, may have these books.)

- If you refer to a color in a Legend or in the text, make sure that it's clear which color you are referring to. Don't just name the color; also show a sample of it right there.

- Blocks of color (or even patterns of black and white), if they are located near text, may distract users.

- Bright colors may seem to "vibrate" when used for text or fine lines.

- Don't use different shades of the same color (such as light blue, medium blue, and dark blue). The colors may be hard to tell apart.

- Avoid several light colors (such as light green and light blue), or several dark colors (such as dark green, dark blue, and black), that may be hard to tell apart.

- *Caution* — Never use *only* color to distinguish important information, especially warnings. Always back it up with something else. For instance, if you put a warning in color, also put it in *bold italic* type. If you use color to distinguish lines on a graph, also label the lines.
  - About 8% of men, and 1% of women, have a problem with color vision.
  - Everyone's color vision is impaired in dim light.
  - Unusual lighting conditions may cause colors to fade, or to seem similar to some other color.
  - The colors may fade over time, especially if the page is exposed to sunlight or other bright light, or if it's handled a lot.
  - The colors will be lost if users photocopy the document.

- If printing on colored paper or card stock, keep in mind that the paper or card stock color may interact with the ink color. Ask the printer to provide you with sample dabs of your choice of inks on the colored paper or card stock.

COMPUTER VERSUS PAPER DOCUMENTS

The trend with some computer systems is to provide user information on the computer system itself, via a *help* command, or via a more elaborate program. This is a good idea, since it ensures that the information will always be available right at the point of use. However, computerized information shouldn't replace paper documents.

Paper documents allow users to underline or otherwise emphasize text that is important to them ... to bend the corners of pages or use paper clips to identify pages important to them ... etc. Also, paper documents usually have better quality pictures, diagrams, etc. than do computer displays. And, of course, users can carry paper documents around with them. (Users may have the capability to print out "help" command information, so that they can have it in paper form. But paper documents can also be made available in convenient booklet or card form, which can't easily be done by individual users with computer printouts.)

According to an article by David A. Schell of the Document Design Center, in *Simply Stated*, February 1986 —
   It takes 20% to 30% longer to read material from computer screens, than from paper.
   Accuracy in detecting errors on computer screens is often lower than it is on paper.
   Problem solving is slower using information read from computer screens than it is from paper. Even people who use a computer all day and thus are used to the computer, do better when reading from paper. And people also do better when reading from paper, even when compared to reading the same material from the best computer screens.
   The article offers some suggestions for improving performance when preparing text to be read on computer screens (similar to suggestions in this book for writing to be read from paper) — Use upper- and lower-case letters, not all upper-case ... Leave the right margin unjustified ... Use medium width lines, not too wide or too narrow ... Don't crowd the information.

FORM

- In addition to a stapled 8½" x 11" document, consider other forms and sizes. It may appear difficult to design a booklet or some other form different from the usual 8½" x 11" document. But it really isn't so difficult, once you become familiar with what needs to be done and the people (editor, artist, printer, etc.) who can help you to do it.

- *Looseleaf* form is easy to update and opens flat. However, it's large and heavy — disadvantages if one has to carry it around ... or use it where there is limited space, as on a crowded desk or at a computer terminal. If you don't intend to keep issuing update pages to the document, some other form is probably better.
  Note — A small size looseleaf book overcomes the disadvantages of size and weight.

- A document with card stock covers can be *stapled* (inexpensive), *plastic comb bound* (moderate), or *wire bound* (somewhat more expensive). There are two types of wire binding. Wire-O type has a line of connected double wires; this may look better and is perhaps more durable, and it is somewhat more expensive. Spiral wire binding (the type found on blank school and pocket notebooks) costs a bit less.
  - This form is more convenient than is looseleaf, especially if comb or wire binding is used, which allows the document to open flat.
  - With comb binding, you can print the title on the spine.
  - You can also print the title on the spine with wire binding if the cover wraps around the wire, but then the book doesn't lie flat. A new technique provides for a flap on the opposite side of the book from the wire binding, so that the title can be printed on the spine and the book can lie flat. The flap can also serve as a bookmark.
  - *Caution* — Ensure that the comb or wire is *big enough,* for the number of pages in the book and the thickness of the paper. If it's too small, the document will be hard to open, and the comb may pop open and allow pages to fall out. The printer or binder takes care of this. Impress on them that you want it done right, and that they will have to do the job over if it isn't done right.
    Most Important — Ask them to send you a few sample bound copies for your approval, before the whole job is bound.

- *Perfect binding* is another form used with card stock covers. The pages are glued together. Paperback books are perfect bound.
  - Doesn't open flat.
  - If poorly glued, the spine may crack, the pages may start falling out, and the whole thing may come apart.

- *Saddlestitching* has two staples in the spine. Many magazines are saddlestitched. (Note that this is different from plain stapling on the edges of the pages, which is known as *sidestitching*.)
  - Opens flat. However, the user has to hold it, or put something on it to keep it lying flat.
  - Often used for booklets.
  - Not recommended for over 100 pages or so.

- *Foldout cards* are popular.
  - Take little space.
  - Open easily.
  - Several pages can be seen at one time.
  - Two cards (one for the front pages, and one for the back pages) can be taped on a wall or desk, thus making it into a poster.
- *Sizes* — 6" x 9" (or some similar size) booklets take less room to use and carry than do 8½" x 11" documents.
- *Summary*

If Your Document	Consider Using					
	Ring Binder	Saddle-stitch Staple	Side-stitch Staple	Plastic Comb	Wire Binding	Perfect Binding
Must be revised —						
Frequently	√					
Rarely		√	√	√	√	√
Must fold —						
So that it can lie flat	√	√		√	√	
For 360-degree page rotation					√	
Contains —						
Few pages (small pamphlet)		√				
Up to 100 pages	√	√	√	√	√	√
More than 100 pages	√		√	√	√	√
Has tabbed dividers	√			√	√	
Must be easy to handle		√		√	√	√
Must be inexpensive		√	√	√		
Needs durable protection	√				√	

- One study showed relative visibility of colors of ink on paper —

Rank	Ink	Paper	
1	Black	Yellow*	(Best)
2	Green	White	
3	Red	White	
4	Blue	White	
5	White	Blue	
6	Black	White	
7	Yellow	White	
8	White	Red	
9	White	Green	
10	White	Black	
11	Red	Yellow*	
12	Green	Red	
13	Red	Green	
14	Blue	Red	(Worst)

  *Also includes cream and ivory

- Another study indicated these colors —

Paper Color	Good Ink Color	Bad Ink Color
White	Black, Blue, Green, Red	Orange, Yellow
Gray	Blue, Black	Orange, Yellow
Yellow	Black, Red	Orange, Green
Red	Black	Green, Blue
Green	Black	Red
Black	White	Red, Orange, Green, Blue
Blue	White	Black

READABILITY

- Ensure that text and diagrams will be clearly legible under the conditions of use. If the document will be used by people in a hurry or under other stress, or under poor lighting or other unfavorable conditions, printing should be larger and clearer than usual.

- Proportional spacing improves legibility. (Each letter gets a different amount of space, according to its size. For instance, *w* gets more space than does *l*. Regular typewriters don't do this.)

- With *justified* right margins, the ends of all lines are lined up *evenly* ... with *un*justified right margins, the ends of lines are *ragged*.
  - Studies indicate that *un*justified right margins are *easier* to read than are justified right margins. *Un*justified right margins do *not* have uneven spaces between words and do *not* have hyphenated words at the ends of lines, as do justified right margins.
  - Also, *un*justified right margins, because each line is a different width, give the eyes better cues when going from one line to the next, so that one is less likely to return to the same line or to skip a line.
  - But on the other hand, justified right margins may appear neater and more professional, probably because people are so used to them in newspapers, magazines, and books.

- Avoid hyphenated words at the ends of sentences. (This isn't possible with justified right margins, unless you are willing to put up with words unevenly spread on the line, and your computer program has an option to do this.)

- Use regular lowercase type (initial capitals only) for text. ALL UPPERCASE IS HARD TO READ.
  - One study estimated that all uppercase slows reading speed from 20% to 50%.

- Don't use *plain italic* type for text or for headings. Italic, especially in uppercase, is hard to read.
  - One study estimates that italic type is 10% to 50% less readable than is regular type. It recommends using boldface type instead of italic type to show emphasis.

- Have all text / charts / examples running the **same way** as the text. Don't make users have to keep turning the document sideways. This is annoying, distracting, and time-consuming. And users may just not bother to read material that is printed sideways.

- Write numbers in **digit**, not in word form ... except, perhaps, at the start of a sentence.
  - Write "There are 3 books" instead of "There are three books." "three" may be the more conventional way, but "3" is clearer.
  - Use commas or spaces if there are many digits (for instance, 3,000,000), unless this would be incorrect.

- In technical material where the distinction is important, type slashes through zeroes (Øs) to clearly distinguish them from uppercase letter Os.

- Have adequate blank space for margins ... and between words, lines, paragraphs, and sections. Keep the spacing uniform.

- Blank space around an item emphasizes its importance. Use this for critical information such as warnings, notices, etc.

- Separate paragraphs by skipping a line between them, whether or not you indent the first line. But don't just indent the first line without skipping a line, as this would cause the text to be too crowded.

- When the final copy is typed, specify lines so that meaningful groups of words or phrases aren't broken at the end of a line (for instance, January 1, 1984 ... page 20 ... Figure 1). **Note** — In the example in this paragraph, "page 20" should appear on 1 line, not on 2 lines as it now is.

- *Type Sizes*
  - *7 point or smaller* — Don't use, except for very detailed charts that won't be used much, and where there is no alternative. In at least one city, it's illegal to have contracts in 7 point or smaller type.
  - *8-9 point* — Use for reference material that will be read for only short periods.
  - *9-11 point* — Use for documents that will be read continuously.
  - *11-12 point* — Use for performance aids that will be used while actually performing a task under adverse lighting or other conditions.
  - *Larger sizes* — Use for headings.
  - *Examples* — Examples of point sizes are shown below. But keep in mind that point sizes vary in different typefaces and on different typesetting machines.

    This is 7 point.

    This is 8 point.

    This is 9 point.

    This is 10 point.

    This is 11 point.

    This is 12 point.

- According to one study, the best line length for most text is 50-70 characters (including spaces), or about 10-12 words. Very *short* lines tire the eyes by making them jump back and forth too much. Very *long* lines tire the eyes by making them stay on one line too long.
  Note — Another study summarizes optimum combinations of type size and line length —

Type Size	*Line Length*
8 points	2-13/16 inches
10 points	3-1/8 inches
12 points	3-13/16 inches

  Note — A point is 1/72 inch or 0.01384 inch.

- Still another study suggests that the ideal combination is 10-point type, 3-1/8 inch line length, and 1 or 2 points of leading (that is, space between lines ... see below).
  - *Note* — A useful item is a Type Size Finder. This is a loupe, similar to that used by watchmakers, but with type sizes engraved on it. Buy it at an Art Supply store or from H. Sherr Engravers, Inc., 7 West 22nd Street, New York, NY 10010. Phone 1-212-242-8630. (An editor, or the Art or Drafting departments in your company, may have one.)

- A useful device for document designers is a circular slide rule for calculating print reduction sizes. Buy it at an Art Supply store. (An editor, or the Art or Drafting departments in your company, may have one.)

- *Leading* is the space between lines of type. The usual thing is to set the leading 1 point larger than the type size.
  - For instance, if type size is 10 point then the leading would be 11 point, and the two would be expressed as 10/11.
  - Very small type and bold type may need extra leading to relieve the density of the text.
  - Long lines may require more leading than short lines, to help the user follow across the long lines.
  - Sans serif type (discussed on the next page) may require more leading than serif type of the same size, because it doesn't have the serifs to carry the horizontal flow of the line.

TYPEFACES

- ***Different typefaces*** (and even the same typeface set by different typesetting machines) may be larger or smaller even though they have the same point size. So you cannot go by point size alone. Also, the shape of the type itself may be different, even though it has the same name. And the leading (see the preceding page) may be different, even though the same points of leading are specified. You must see ***actual samples*** of the type, set on the typesetting machine to be used, before you make your choice of type styles, sizes, and leading. For example, here are two examples of different type styles set in 10 point and in 14 point. You can see the difference in sizes.

This is an example of 10 point Souvenir type.

This is an example of 10 point Avant Garde type.

This is an example of 14-point Helvetica.

This is an example of 14-point Garamond.

- ***Serif Vs. Sans Serif Type*** —
  - ***Serif*** type styles have short strokes projecting from tops and bottoms of letters. Serifs may enhance the horizontal flow of lines, perhaps making lines easier to read ... may make individual letters easier to distinguish ... and are considered to be more traditional and formal. ***Note*** — This sentence is typeset in Times Roman, a serif type style.
  - ***Sans serif*** type styles don't have serifs. They have a clean, modern, informal look. The simpler look may make these styles easier to read.
    Note — This paragraph, and the rest of this book, is typeset in Helvetica, a sans serif type style.

BIAS IN CONCEPT AND COVERAGE — OMISSION

Check the descriptive and illustrative material — the examples used to illustrate concepts, and the descriptions of processes, social structures, and typical situations. Are women simply ignored? Are they treated as exceptional cases or, on the other hand, as part of the landscape or the baggage? Are the subjects of studies all male? Is the work of women scholars cited? Certain subjects — history, the sciences, and business — are special candidates for careful scrutiny. The argument usually advanced — that there are no women involved or that women did not play certain roles in certain periods or cultures — does not justify ignoring women altogether or mentioning them only as auxiliaries or oddities.

Unbiased	*Biased*
Pioneer families crossed the desert carrying all their possessions.	The pioneers crossed the desert with their women, children, and possessions.
Slave families were allowed to stay together. *or* Married slaves were allowed to live with their families.	The slaves were allowed to marry and to have their wives and children with them.
Marie Curie discovered radium.	Radium was discovered by a woman, Marie Curie.
When setting up an experiment, a researcher must check for sampling error.	When setting up his experiment, the researcher must check his sample for error.
As knowledge of the physical world increased, old ideas and traditions were examined with a more critical eye. *or* ... people began to examine	As knowledge of the physical world increased, men began to examine long-held ideas and traditions with a more critical eye.

BIAS IN CONCEPT AND COVERAGE
— EQUAL TREATMENT

Check the use of adjectives and modifiers. Do those used for women consistently create a negative impression or betray a patronizing attitude? Are women mentioned consistently as an afterthought? Does the inclusion of women seem like a conscious effort or a concession on the part of the author? An attempt to be trendy or up-to-date? Are women consistently described in physical or sexual terms that are never used when describing men? Is a women's marital status always mentioned even though the context does not require it?

Unbiased	*Biased*
The exhausted pioneers .... *or* The exhausted men and women	The poor women could no longer go on; the exhausted men ....
The women painters of this period were ... *or* worked primarily in ... *or* Among the painters of this period, X, Y, and Z [both men and women] worked primarily in	There were also some women painters in this period, most of them daughters or wives of painters.
She ran the business efficiently.	Though a woman, she ran the business efficiently.
Jan Acton is Joe Granger's assistant.	Mrs. Acton, a statuesque blonde, is Joe Granger's assistant.
The girls played with the boys. *or* The children played. *or* The little girls played with the little boys.	The little girls played with the boys.
The line manager and his secretary were both upset by the mistake.	The line manager was angry; his secretary was upset.
All the young people of the village took part in the festival.	All the strong young men of the village took part in the festival, as did the young girls.

BIAS IN CONCEPT AND COVERAGE
— STEREOTYPING

Check the portrayal of roles, the description of jobs and skills, the treatment of life styles and life situations. Are people treated as human, or are all the portrayals done in male or female terms? Does the reader get the impression that only men do X and only women do Y? Are men portrayed one way and women another? Are all people in positions of authority or trust (the therapist, the politician, the scientist, the philosopher, the leader, the historian) male? In an education text, are all the teachers female and all the professors and administrators male? How is the family described and analyzed? Do "mommies" always stay at home? Does the author imply they should? What are the role models for children? How are recent changes in the family power structure treated? In a business text, are all the executives male and all the secretaries and assistants female? Can the reader instantly infer that all the participants in a meeting or conference are assumed to be male? Do examples of human behavior always reinforce the stereotyped idea that women and men are totally different kinds of creatures?

Unbiased	**Biased**
She was actively interested in sports as a child.	As a child, she was a tomboy; sports and not dolls were her main interest.
Current tax regulations allow a head of household to deduct for the support of a spouse and children.	Current tax regulations allow a head of household to deduct for the support of a wife and children.
The line manager is responsible for the productivity of the department; the supervisors, for that of the workers on the line.	The line manager is responsible for the productivity of his department; his foremen, for the day-to-day work of the girls on the line.
The secretary brought the boss coffee.	The secretary brought her boss his coffee.
The teacher should prepare a lesson plan well in advance of the day the material will be taught. *or* The teacher must be sure his or her lesson plans	The teacher must be sure her lesson plans are done well in advance of the day she plans to teach the material.

Completing this "awareness checklist" should give an editor (or an author) a good idea of whether or not a particular manuscript needs more than the adjustment of pronouns or changes in language to make it unbiased or non-sexist. At this point the editor or the copy editor can evaluate the scope of the problem and make recommendations for substantive work in addition to changes in language and expression. It is the overall presentation, not so much the occasional lapse in language, that can give a book a bias the author may not have intended.

BIAS IN LANGUAGE AND EXPRESSION — PRONOUNS

The use of *he, his, him* to denote any person or a person is the most common problem in editing language for bias, simply because English has no neutral pronoun in the singular. If there is no way to reword a passage or a sentence to avoid unnecessary pronouns or to change to the plural, the best current solution is to use *he* or *she*, *his* or *her*. Coined terms, such as *(s)he* or *she/he*, should be avoided; they are usually distracting to the reader and annoying to the author. In citing examples, individuals in the examples may sometimes be male, sometimes female. If this alternative is chosen, avoid stereotyping male and female roles.

None of these suggestions — removing pronouns, changing to the plural, alternating examples, or substituting *he* or *she* — should be followed blindly as immutable rules. Context and clarity of expression are important considerations in a text and should not be sacrificed merely to ensure that every pronoun has been changed. The constant use of *her* or *she* leads to clumsy, repetitious phrases and sentence structure. A change to the plural may be wrong in a given context — for example, when the discussion is of one-to-one relationships such as that between parent and child. Alternating the *he* and *she* examples does not work well in many contexts. Both author and editor need to use a variety of approaches in sensitive, appropriate ways. It is advisable for the author to include a note in the preface to the book explaining what approach has been taken to avoid stereotyping and sexism.

Unbiased	**Biased**
Facial expression does not always indicate a person's true feelings.	A person's facial expression does not always reveal his true feelings.
Sometimes a doctor will see patients only in a hospital.	Sometimes a doctor will see his patients only in a hospital.
The clinician must take accurate and careful measurements.	The clinician must take his measurements accurately and carefully.
Most children do their homework right after school.	The typical child does his homework right after school.
A good lawyer will see that his or her clients are aware of their rights.	A good lawyer will see that his clients are aware of their rights.

Parallel Treatment	*Non-Parallel Treatment*
The men in the office took the women to lunch.	The men in the office took the girls to lunch.
This is my secretary, Alice Smith and my aide, Jack Green.	This is my secretary, Mrs. Smith and my aide, Jack Green.
Dr. and Mrs. Jones *or* Jack and Diane Jones	Dr. Jones and his wife Diane
husband and wife	man and wife
student	co-ed
college men and women	college men and girls
the men and the women *or* the gentlemen and the ladies	the men and the ladies
at a meeting between President Nixon and Prime Minister Gandhi [*or* Richard Nixon and Indira Gandhi, Mr. Nixon and Mrs. Gandhi]	at a meeting between President Nixon and Indira Gandhi [*or* Mrs. Gandhi]

Avoid using cliches such as the following —

the woman driver	boys' night out
the nagging mother-in-law	dizzy blonde
the little woman	catty women
the henpecked husband	female gossip
gal Friday	man-size job

BIAS IN LANGUAGE AND EXPRESSION — OCCUPATIONS AND TITLES

Naming a person's occupation has been an editorial problem, simply because so many job titles and occupations were themselves gender-linked terms. Many alternatives are now available, so that it is usually easy to use descriptive words that can apply to any person, whether male or female. Unnecessary gender identification can also be deleted.

Unbiased	Biased
actor	actress
author	authoress
businessperson, executive, manager	businessman
chair, chairperson	chairman
household worker, cleaner	cleaning lady
member of Congress, representative	congressman
supervisor	foreman
firefighter	fireman
servant	houseboy
homemaker, consumer	housewife
mail carrier, letter carrier	mailman
police officer	policeman, policewoman
sales representative, salesperson	salesman, saleswoman
flight attendant	stewardess
doctor	woman doctor
nurse	male nurse

MORE SUGGESTIONS

- When it is not relevant, do not mention characteristics as sex, marital status, race, age, physical appearance, etc.

- The use of *they* as a singular generic pronoun is grammatically correct and fully provided for in the Oxford English Dictionary. "If someone wants to go to college, here's what they should know," reads a line in a New York Times advertisement. Another example is, "When a soldier is firing the 105mm Howitzer, they should wear ear protection." No forcing of the language is necessary, we are simply using a form that has recently fallen into disuse.

- Ask colleagues of both sexes to review what you write.

- Although you may not and need not agree with a female ... or male ... interpretation of events, you should include the female or male perspective in any review of the possible determinants of behavior by groups or individuals.

- Women authors are often under-represented in text citations in proportion to the number of eminent women in a particular field. Include references to research done by women when it is relevant to the topic of discussion.

- Use non-sex-typed examples. Use female as well as male names for prototype doctors, pilots, or mechanics. Use male as well as female names for prototype child caretakers, homemakers, or clerical workers. Also, any discussion of occupations and career choices should not imply that members of only one sex have the desire for, or access to, a particular goal in life.

- Give attention to individuals depicted in illustrations and photographs so that stereotyped views of "typical male" and "typical female" activities are not reinforced.

- Have an independent reviewer or panel analyze your survey to assure that it is free of ethnic or sex role stereotyping, or racial bias. Your company's editorial or personnel department may be helpful.

GRAMMAR

ACKNOWLEDGEMENT

Material in this chapter on **GRAMMAR** is summarized and adapted from **Guide for Air Force Writing** (AFM 10-4), U.S. Government Printing Office, Washington, D.C., 1960; and **The Language of Audit Reports** by Laura Grace Hunter, Washington, D.C., U.S. Government Printing Office, 1957.

MODIFIERS

- **Modifiers** — Misplaced modifiers (see illustrations below) make it easy for the user to misunderstand the meaning of sentences, sometimes with dire results. **Keep your modifiers close to the words they modify**.

- **Dangling Modifier** — When a word or phrase seems to modify another word which it cannot logically modify, it has been left dangling. Usually it will be a phrase beginning the sentence. From its position we expect it to modify the subject. But the connection is illogical.
  - **Say** "To make a climbing turn, open the throttle wider."
    Not "Making a climbing turn, the throttle is opened wider."

- **Misplaced Modifier** — Sometimes we widely separate a modifier from its modified word and confuse the user.
  - **Say** "It was impossible in the dark to find the book I had been reading."
    Not "It was impossible to find the book that I had been reading in the dark."

- **Squinting Modifier** — The modifier may be placed so that it could logically relate to either of two words. This may confuse the user.
  - **Say** "The electrician said he would repair the stove on Wednesday."
    Not "The electrician said Wednesday he would repair the stove."

INDEFINITE REFERENCE WORDS

- We use words or pronouns such as "the latter," "the former," "this," to refer to something we have previously mentioned. This reference must be clear to the user.
  - *Say* "The commander told the executive to handle all personnel assignments." *or* "The commander told the executive that she, the commander, would handle all personnel assignments."
    Not "The commander told the executive that she would handle all personnel assignments."

SINGULAR AND PLURAL

- Don't carelessly carry along a singular verb to a second subject in the plural, or a plural verb to a second subject in the singular.
  - *Say* "A payroll *is* prepared and checks *are* drawn."
    Not "A payroll *is* prepared and checks drawn."
  - *Say* "Storage facilities *are* inadequate and their location *is* undesirable."
    Not "Storage facilities *are* inadequate and their location undesirable."

- The verb agrees with its subject, regardless of other words or expressions that come between them.
  - *Say* "The *record* of these valuations and changes *is* kept."
    Not "The *record* ... are kept."

- "Each," "every," "either," and "neither" always take singular verbs.
  - *Say* "Each of the guests *was* served."
    "Every employee of these companies *is* required ...."
    "Either of the methods *is* acceptable."
    "Neither of the departments *is* responsible."

GERUNDS

- A gerund is the "ing" form of a verb used as a noun. It takes a possessive qualifier. It is a complex and awkward form. The best thing to do is to reword the sentence to avoid using it. Often it is difficult to recognize a gerund. There is one simple test to find out whether an "ing" word is a gerund requiring a possessive modifier: try substituting the pronoun. If the pronoun must be a possessive pronoun, then the noun it replaced should be in the possessive form.
  - "This offer resulted in the *office receiving* many applications." — Substitute the pronoun for "office." "This offer resulted in (*it, its?*) receiving many applications." Obviously *its* is correct, so we should use the possessive form of "office." "This offer resulted in the *office's receiving* many applications." But it would be simpler just to say: "As a result of this offer, the office received many applications."

IDIOMS

- An idiom is a word usage that depends on the context to be understood.
  - Agree *to* a proposal; agree *with* a person.
  - Argue *with* a person; argue *for, against,* or *about* (but not *on*) a proposal.
  - Blame *for* (not *on*).
  - Compare *to* a standard or base previously set up; compare *with* another item.
  - Differ *with* a person; differ *from* anything else.
  - Independent *of* (not *from*).
  - Merge *into* or *in* (but not *with*).
  - Plan *to* or *for* (not *on*).
  - Wait *on* a customer; wait *for* a person or thing; wait *at* a place.

- Certain verbs are idiomatically followed by *to be*.
  - Net results continued *to be* profitable.
  - These limitations appear *to be* related to the programs.

- Verbs of saying, thinking, feeling, believing are idiomatically followed by *that*.
  - We feel *that* the statements ....
  - The manager stated *that* this action ....

SPLIT INFINITIVES

- An infinitive is a verb form that contains the word **to**.
  - to eat
  - to be going

  An infinitive is said to be split when a word or phrase occurs between **to** and the verb.
  - to immediately eat
  - to soon be going

  Split infinitives are objectionable only if they are split without reason. If it makes a sentence stronger and more natural to split the infinitive, split it!
  - The board should have the authority necessary **to effectively direct** the affairs ...

SPLIT PARTICIPLES

- Split participles are no more objectionable than split infinitives.
  - Every organization **should carefully plan** ahead ....
  - This position **could best be** filled by ....

END THAT SENTENCE WITH A PREPOSITION!

- A prepositional ending has never been incorrect. Often a preposition is a good word to end a sentence **with**.

MISCELLANEOUS

- The same words have a hyphen between them in some contexts where they are used as one word, but not in others where they are used as two words.
  - The on-line system was installed.
  - I wrote the book on line.
  - The out-of-date magazine is here.
  - The magazine is out of date.

- Use **which** in **non-restrictive** clauses (clauses that don't change the meaning of the sentence). Use **that** in **restrictive** clauses (clauses that are essential to the meaning of the sentence).
  - The boat, which was big, was at the dock.
  - Boats that are big must have special lights.

- **Who** refers to people. **That** and **which** refer to animals and things.
  - The people who came to the party ....
  - The box that is on the table ....

- **i.e.** means **that is**.
  - The problem, i.e., the malfunctioning clock, was attended to.

- **e.g.** means **for example**. Don't use "etc." at the end of an **e.g.** list, because the **e.g.** means that you are just giving a few examples and that there are more instances of the same category.
  - Fruits, e.g., apples and pears, are popular.

- **etc.** means **and the rest**.
  - They visited many cities, such as New York, Chicago, Los Angeles, etc.

- **sic** indicates that you have recognized an error in the preceding word(s) of a quotation, but that you haven't changed the word(s) because it is in a quotation. (Enclose the word **sic** in brackets, to show that it isn't part of the quotation.)
  - "It's [sic] counterpart ...."
    In this example, there should **not** be an apostrophe in **It's**. It should be **Its**. (The apostrophe is used when **It's** is a contraction of **It is** ... e.g., "It's sunny outside." **Its** is used in the above example in the possessive case.)

PUNCTUATION

ACKNOWLEDGEMENT

Material in this chapter on **PUNCTUATION** is summarized and adapted from **U. S. Government Correspondence Manual**, Washington, D.C., U.S. Government Printing Office, 1968.

Note — The above government publication is not copyrighted. However, the summarized and adapted material in this book is covered by the copyright of this book, and may not be used unless you get the written permission of the publisher of this book.

ACRONYMS

- Show the plural by adding an **s**.
  - CRTs

- To make an acronym possessive, add **'s**.
  - The CRT's screen

- If the line is all capitals (as in a heading), add an **'S**.
  - THE CRT'S SCREEN

- Precede an acronym with —
  - **a** — If it begins with a consonant sound.
    (**a** CRT terminal)
  - **an** — If it begins with a vowel sound.
    (**an** MBA degree)

164

APOSTROPHE

- Use to indicate contractions or omitted letters.
  - I've
  - It's (it is)
  - TV'ers

- Use to indicate the coined plurals of letters, figures, and symbols.
  - three R's
  - 5's and 7's
  - +'s

- Use to show possession. Add *'s* when the noun doesn't end with an *s* sound. Add *'s*, or only the apostrophe, to a noun that ends with an *s* sound (either way is ok).
  - officer's
  - hostess'
  - Co.'s
  - Cos.'
  - Jones's
  - Jones'
  - Joneses'

- To show possession in compound nouns, add the apostrophe or *'s* to the final word.
  - secretary-treasurer's

- To show joint possession in nouns in a series, add the apostrophe or *'s* to the last noun.
  - soldiers and sailors' home

- To show separate possession in nouns in a series, add the apostrophe or *'s* to each noun.
  - Diane's, John's, Thomas', and Mary's

- To show possession in indefinite pronouns, add the apostrophe or *'s* to the last component of the pronoun.
  - someone's desk
  - somebody else's books
  - others' homes

- Do *not* use to form the possessive of personal pronouns.
  - theirs
  - yours
  - hers
  - its

- Do **not** use to form the plural of spelled-out numbers, of words referred to as words, and of words already containing an apostrophe. Add **'s**, however, if it makes the plural easier to read.
  - twos and threes
  - yeses and noes
  - which's and that's
  - ifs, ands, and buts
  - do's and don'ts

- Do **not** use to follow names of countries and other organized bodies ending in **s**, or after words more descriptive than possessive (not indicating personal possession), except when the plural doesn't end in **s**.
  - United States control
  - merchants exchange
  - children's library

COMMA

- Use to separate words or figures that might otherwise be misunderstood or misread.
  - Instead of hundreds, thousands came.
  - Out of each 20, 10 are accepted.

- Use to set off introductory or explanatory words that precede, break, or follow a short direct quotation. The comma isn't needed if a *?* or *!* is already part of the quoted matter.
  - I said, "Don't you understand?"
  - "Why?" they said.

- Use to indicate the omission of understood word(s).
  - Then they were enthusiastic; now, indifferent.

- Use to separate a series of modifiers of equal rank.
  - It is a young, eager, intelligent group. **but** That is a clever young person. (No comma when the final modifier is considered part of the noun modified.)

- Use to follow each of the members within a series of three or more, when the last two members are joined by "and," "or," or "nor."
  - horses, mules, and cattle
  - by the bolt, by the yard, or in remnants
  - neither snow, rain, nor heat

- Use to separate an introductory phrase from the subject it modifies.
  - Beset by the enemy, they retreated.

- Use before and after *Jr., Sr.,* academic degrees, and names of states preceded by names of cities, within a sentence.
  - H. Smith, Jr., Chairperson
  - Washington, D.C., schools

- Use to set off parenthetic words, phrases, or clauses.
  - The new model, developed by the XYZ project, was first used on that project.
  - **but** The person who gave that talk is undoubtedly a good speaker. (No comma necessary, since the clause "who gave that talk" is essential to identify the person.)

- Use to set off words or phrases in apposition or in contrast.
  - Pat Smith, attorney for the plaintiff, asked for a delay.
  - You will need work, not words.

- Use to separate the clauses of a compound sentence if they are joined by a simple conjunction such as "or," "nor," "and," or "but."
  - The United States will not be an aggressor, nor will it tolerate aggression by other countries.

- Use to set off a noun or phrase in direct address.
  - Ms. President, the motion has carried.

- Use to separate the title of an official and the name of the organization, in the absence of the words "of" or "of the."
  - Chief, Insurance Branch
  - Chairperson, Committee on Appropriations

- Use to separate thousands, millions, etc., in numbers of four or more digits.
  - 4,230
  - 50,491
  - 1,000,000

- Use to set off the year when it follows the day of the month in a specific date within a sentence.
  - The reported dates of September 11, 1943, to June 12, 1955, were erroneous.

- Use to separate a city and state.
  - Cleveland, Ohio
  - Washington, D.C.

- Do **not** use to separate the month and year in a date.
  - Production for June 1955

- Do *not* use to separate units of numbers in built-up fractions, decimals, page numbers, serial numbers (except patent numbers), telephone numbers, and street addresses.
  - 1/2500
  - Motor No. 189463
  - 1.9047
  - 639-3201
  - page 2632
  - 1727-1731 Broad Street
  - 1450 kilocycles, 1100 meters (no comma unless more than four digits, radio only)

- Do *not* use to precede an ampersand (&) or a dash.
  - Greene, Wilson & Co. (except in indexes: Jones, A.H., & Sons)
  - There are other factors — time, cost, and transportation — but quality is the most important.

- Do *not* use to separate two nouns one of which identifies the other.
  - The booklet "Infant Care"
  - Wilson's boat *The Maria*

- Do *not* use to separate the name and the number of an organization.
  - Western Legion Post No. 12

- Do *not* use as the sole connection between two independent clauses (this is known as a comma fault). Use a period or a semicolon instead of the comma. Or put an "and" after the comma.
  - *Say* "The picnic is an annual event. This year it will be held at the park."
  - *Not* "The picnic is an annual event, this year it will be held at the park."

DASH

- A dash may or may not be preceded and followed by a space. I feel that it looks better if it is preceded and followed by a space, as in the examples below (and elsewhere in this book).

- Use to mark a sudden break or abrupt change in thought.
  - She said — and no one contradicted her — "The battle is won."

- Use to indicate an interruption or an unfinished word or sentence.
  - He said, "Give me lib —"
  - Q. Did you see — ?

- Use instead of commas or parentheses, if the meaning is clarified by the dash.
  - These are shore deposits — gravel, sand, and clay — but marine sediments underlie them.

- Use to introduce a final clause that summarizes a series of ideas.
  - Freedom of speech, freedom of worship, freedom from want, freedom from fear — these are the fundamentals of moral world order.

- Use to follow an introductory phrase leading into two or more successive lines and indicating repetition of that phrase.
  - I recommend —
    That we accept the rules
    That we publish them

- Use instead of a colon when a question mark closes the preceding idea.
  - How can you explain this? — "Fee paid, $5."

- Use to precede a credit line or signature.
  - Still achieving, still pursuing,
    Learn to labor and to wait.
    — Henry Wadsworth Longfellow

ELLIPSIS

- Ellipsis consists of 3 dots, with a space before the 1st dot and after the last dot. If it occurs at the end of the sentence, add a 4th dot.

- Use to indicate an omission within a quotation.
  - They said, "One two three ... seven eight."
  - They said, "One two three ...."

- Use instead of a dash if you want to. (See **DASH** on the preceding page.)
  - She said ... and no one contradicted her ... "The battle is won."

EXCLAMATION POINT

- Use to mark surprise, incredulity, admiration, appeal, or other strong emotion, which may be expressed even in a declarative or interrogative sentence.
  - How beautiful!
  - "Great!" they exclaimed.
  - Who shouted, "All aboard!" (Question mark omitted)
  - O Lord, save Thy people!

HYPHEN

- Use to connect the elements of certain compound words.
  - mother-in-law
  - self-control
  - ex-governor
  - walkie-talkie

- Use to indicate continuation of a word divided at the end of a line.
  - This line is cont-
    inued on the next line.

- Use to separate the letters of a word which is spelled out for emphasis.
  - d-o-l-l-a-r-s

- Use to modify a letter or number.
  - 10-cent charge
  - 32-bit computer

- Use to avoid doubling a letter.
  - re-evaluate

- Use if the main word begins with a capital letter.
  - post-World War II

- Use to avoid awkward pronunciations or ambiguity.
  - co-worker
  - re-read

- Use after a series of words with a common base.
  - small- and medium-sized companies

NUMBERS

- Spell out numbers at the beginning of a sentence. Also spell out numbers under 10, except when expressing time, money, and measurement. *Note* — In my opinion, it may be clearer to put all numbers in number form — e.g., 4 — especially in technical material.
  - Four days ago ....
  - There are two items ....
  - There are 15 items ....
  - There are 2 items ....

- For numbers less than one, precede the decimal point with a zero to emphasize the decimal point.
  - 0.03
  - 0.4

- Prefer Arabic numerals to Roman numerals.
  - *Say* "10"
    Not "X"

- Except in legal documents, avoid repeating a number which has been spelled out.
  - *Say* "There are two items ...."
    Not "There are two (2) items ...."

PARENTHESES

- Use to set off matter not part of the main statement or not a grammatical element of the sentence, yet important enough to be included.
  - Kelley (to the chairperson).
  - Q. (Continuing.)
    A. (Reads:)
  - The result (see figure 2) is most surprising.

- Use to enclose a parenthetic clause where the interruption is too great to be indicated by commas.
  - You can find it neither in French dictionaries (at any rate, not in Littré) nor in English dictionaries.

- Use to enclose an explanatory word that is not part of the statement.
  - The Erie (Pa.) Ledger *but* The Ledger of Erie, Pa.

- Use to enclose letters or numbers designating items in a series, either at the beginning of paragraphs or within a paragraph.
  - You will observe that the sword is (1) old fashioned, (2) still sharp, and (3) unusually light for its size.

- Use to enclose a reference at the end of a sentence. Unless the reference is a complete sentence, place the period after the parenthesis closing the reference. If the sentence contains more than one parenthetic reference, the parenthesis closing the reference at the end of the sentence is placed before the period.
  - The specimen exhibits both phases (pl. 14, A, B).
  - The individual cavities show great variation. (see pl. 4.)
  - This sandstone (see pl. 6) occurs in every county of the State (see pl. 1).

- Put an open parenthesis at the beginning of each paragraph, but a close parenthesis only at the close of the last paragraph, when extensive material is enclosed.
  - (This is paragraph 1.

    ...

    (This is paragraph 2.

    ...

    (This the last paragraph.)

PERIOD

- Use to end a declarative sentence that is not exclamatory, and to end an imperative sentence.
  - They work for Johnson, Inc.
  - Do not be late.

- Use to end an indirect question or a question intended as a suggestion and not requiring an answer.
  - Tell me how the rocket was launched.
  - May we hear from you soon.

- Use three periods to indicate omission within a sentence; at the end of a sentence, use four periods.
  - He called ... and left ....

- Use to follow abbreviations unless by usage the period is omitted.
  - gal.
  - NE.
  - qt.
  - *but*
  - HEW
  - USDA

QUESTION MARK

- Use to indicate a direct query, even if not in the form of a question.
  - Did they do it?
  - They did what?
  - Can the money be raised? is the question.
  - Who asked, "Why?" [Note single ?]

- Use to express more than one query in the same sentence.
  - Can they do it? or you? or anyone?

- Use to express doubt.
  - They said the table was 8(?) feet tall.

- Do *not* use if the sentence is really a request rather than a question.
  - Will you enter my subscription to your magazine.

QUOTATION MARKS

- Use to enclose a direct quotation. Single quotation marks are used to enclose a quotation within a quotation.
  - The answer is "No."
  - "Your order has been received," they wrote.
  - They said, "Pat said 'No.'"
  - "Pat," said Diane, "why are you late?"
  - "The equipment will be forwarded promptly."

- Use to enclose any matter following the terms "entitled," "the word," "the term," "marked," "endorsed," or "signed." Do *not* use to enclose expressions following the terms "known as," "called," "so-called," ..., unless such expressions are misnomers or slang.
  - Congress passed the act entitled "An act ...."
  - It was signed "Mary."
  - After the word "treaty," insert a comma.
  - The so-called investigating body.

- Use to enclose misnomers, slang expressions, nicknames, or ordinary words used in an arbitrary way. But don't overdo this usage; it could result in an affected writing style.
  - Some "antiques" might better be described as junk.

- Limit quotation marks, if possible, to three sets (double, single, double).
  - "The question is, in effect, 'Can a person who obtained a certificate of naturalization by fraud be considered a "bona fide" citizen of the United States?'"

- Type the comma and the final period inside the quotation marks. Other punctuation marks are placed inside only if they are a part of the quoted matter.
  - "The President," they said, "will veto it."
  - The conductor shouted, "All aboard!"
  - Is this what we call a "Correspondex"?
  - "Have you an application form?"
  - Who asked, "Why?"

- Put open quotes at the beginning of each paragraph, but close quotes only at the close of the last paragraph, when extensive material is enclosed.
  - "This is paragraph 1.

    ...

    "This is paragraph 2.

    ...

    "This is the last paragraph."

SEMICOLON

- Use to separate independent clauses not joined by a conjunction, or joined by a conjunctive adverb such as "hence," "therefore," "however," "moreover," ....
  - The report is not ready today; it may be completed by Friday.
  - The allotment has been transferred; hence, construction must be delayed.

- Use to separate two or more phrases or clauses with internal punctuation.
  - If you want your writing to be worthwhile, give it unity; if you want it to be easy to read, give it coherence; and, if you want it to be interesting, give it emphasis.

- Use to separate statements that are too closely related in meaning to be written as separate sentences.
  - No; we receive one-third.
  - War is destructive; peace, constructive.

- Use to precede words or abbreviations which introduce a summary or explanation of what has gone before in the sentence.
  - A writer should adopt a definite arrangement of material; for example, arrangement by time sequence, by order of importance, or by subject classification.
  - The industry is related to groups that produce finished goods; e.g., electrical machinery and transportation equipment.

MORE PUNCTUATION RULES

- Periods and commas always go inside double quote marks. The only allowable exception is a single character in quotes. **Note** — If you are specifying information to be entered into a computer, you had best put the period **outside** the close quotes, or users might think that the period is part of the command (see 3rd example below). Better still, don't use quotes at all — use a dash ... skip a space ... specify the information (perhaps in italics or bold italics) ... and then skip a space before the period (see 4th example below). In England, they **always** put the periods and commas outside the close quotes.
  - "We want to go to the fair," they said.
  - Use a tilde "˜".
  - Enter into the computer, "go".
  - Enter into the computer — **go** .

- Semicolons always go outside double quotes.
  - They knew what was meant by "hardcopy"; they didn't know about "software."

- Question marks and exclamation marks go inside or outside double quote marks depending on the sentence sense.
  - "Where are you going?" they asked.
  - What is meant by the word "firmware"?

- When a quote ends with a question mark that ends a clause, and a comma would normally appear at the end of the clause, it is standard to leave the comma out.
  - "Where are you going?" they asked.

- When using single quote marks instead of double quote marks, the same rules apply. (Single quotes are considered incorrect, except inside a quotation enclosed in double quotes.)

- When a sentence is enclosed in parentheses, the period goes inside the closing parenthesis.
  - (This is a sentence.)

- If the words inside parentheses do not constitute a sentence, but are at the end of the sentence, the period goes after the closing parenthesis.
  - This is a sentence (but not this).

- No commas, semicolons, or colons should appear before a left parenthesis. If such punctuation is needed, it is placed after the phrase in parentheses.
  - After eating salt (sodium chloride), they threw up.

- Dashes never occur next to commas, semicolons, or parentheses. The following example is **wrong**. Use either the commas or the dashes, but not both.
  - Before World War I, — but not afterwards, — Iceland was a part of Denmark.

ACTION VERBS/ ABSTRACT NOUNS

ACKNOWLEDGEMENT

Material in this chapter on
ACTION VERBS/ABSTRACT NOUNS is from
The Language of Audit Reports by Laura Grace Hunter,
U.S. Government Printing Office, Washington, D.C., 1957.
(See page 46 for a discussion of
Action Verbs/Abstract Nouns.)

Action Verb	Abstract Noun	Action Verb	Abstract Noun
	(some of the *ation* family)		
allocate	allocation	inform	information
anticipate	anticipation	justify	justification
apply	application	limit	limitation
classify	classification	minimize	minimization
compile	compilation	negotiate	negotiation
compute	computation	observe	observation
condemn	condemnation	operate	operation
confirm	confirmation	prepare	preparation
consider	consideration	preserve	preservation
create	creation	reactivate	reactivation
delegate	delegation	reconcile	reconciliation
depreciate	depreciation	record	recordation
designate	designation	regulate	regulation
determine	determination	relate	relation
document	documentation	represent	representation
eliminate	elimination	segregate	segregation
evaluate	evaluation	standardize	standardization
examine	examination	stipulate	stipulation
explore	exploration	tax	taxation
formulate	formulation	utilize	utilization
implement	implementation	verify	verification

Action Verb	Abstract Noun	Action Verb	Abstract Noun

(some of the *ment* family)

Action Verb	Abstract Noun	Action Verb	Abstract Noun
abandon	abandonment	establish	establishment
adjust	adjustment	improve	improvement
agree	agreement	invest	investment
appraise	appraisement	manage	management
arrange	arrangement	pay	payment
ascertain	ascertainment	procure	procurement
assess	assessment	recruit	recruitment
assign	assignment	reimburse	reimbursement
demolish	demolishment	relinquish	relinquishment
develop	development	replace	replacement
disburse	disbursement	require	requirement
enact	enactment	retire	retirement
enforce	enforcement	settle	settlement
entitle	entitlement	treat	treatment

(some others)

Action Verb	Abstract Noun	Action Verb	Abstract Noun
absorb	absorption	execute	execution
accept	acceptance	exhaust	exhaustion
acquire	acquisition	expend	expenditure
adopt	adoption	include	inclusion
assume	assumption	inspect	inspection
collect	collection	instruct	instruction
compare	comparison	intend	intention
complete	completion	issue	issuance
comply	compliance	maintain	maintenance
construct	construction	perform	performance
consume	consumption	prefer	preference
contribute	contribution	produce	production
convert	conversion	recognize	recognition
convey	conveyance	reduce	reduction
deduct	deduction	refer	reference
deliver	delivery	remit	remittance
diminish	diminution	remove	removal
direct	direction	render	rendition
disclose	disclosure	renew	renewal
dispose	disposition	restrict	restriction
dissolve	dissolution	retain	retention
distribute	distribution	revise	revision
divert	diversion	supervise	supervision
emerge	emergence	transact	transaction

REFERENCES

For those interested in some basic books on writing, I suggest the following. In addition, see the Index (next page) for books that I have referenced. Also, I would suggest that you go to a large library or, even better, a large bookstore (which is more likely to have the most popular and newest books) and see which could be of use to you. Also, ask your friends, professors (if you are in college), and others for their recommendations.

- *Simply Stated* — This is an interesting and very useful newsletter published by the Document Design Center (see page 36 for address and phone). It reports on developments in the writing field, including the Plain English movement (to rewrite contracts, laws, etc., in more readable language), has articles on writing, and reviews of new books on writing. The Center also publishes some useful books on writing. There is no charge for the newsletter.

- *Computer Programs* — If you use a computer to write, there are probably dictionaries, thesauruses, and perhaps writing style programs for your computer that you might want to look into.

- *The Elements of Style* by William Strunk Jr. and E. B. White, Macmillan Publishing Co., Inc., 1979. A concise, basic book that has stood the test of time in helping many people.

- *Webster's New International Dictionary*, Merriam-Webster, Inc. An excellent unabridged dictionary.

- *Webster's New Collegiate Dictionary*, Merriam-Webster, Inc.

- *Oxford American Dictionary*, Oxford University Press.

- *The American Heritage Dictionary of the English Language*, Houghton Mifflin Company.

- *Roget's II The New Thesaurus*, by the editors of *The American Heritage Dictionary*, Houghton Mifflin Company.

- *Webster's New Dictionary of Synonyms*, G. and C. Merriam Company.

INDEX

188

CONTENTS

Disney's
Cooking
with
Mickey
& Friends

Healthy Recipes from Your Favorite Disney Characters

Pat Baird

Disney
PRESS

New York

Contents

A Word from Mickey

Hi, boys and girls!

Welcome to the kitchen. I bet you're surprised to see me here. Well, I'm not alone. All your other Disney pals are here, too. Most of the time you see us in all sorts of adventures. We run, jump, and play. We've all got lots of energy. That's because we like to eat, and we eat well. So we thought we would share some of our favorite recipes with you.

We picked recipes that everyone of all ages will like to eat. Sometimes you might see a recipe or an ingredient you haven't eaten before. Try it! Part of the fun of eating is tasting new food. The other part of eating is that it helps us to think better, grow taller and stronger, and even play better.

How do you know what to eat and how much? Think *variety*. This involves eating a few different foods from all the food groups every day.

Whole grain cereals, breads, and grains	**6 or more servings**
Vegetables	**3 or more servings**
Low-fat milk, cheese, yogurt, and other dairy products	**3 or more servings**
Lean meats, poultry (such as chicken and turkey), eggs, beans, nuts, or seeds	**2 servings**
Fruits	**2 or more servings**

The meat department of your local supermarket certainly offers a great deal of variety—so which are the lean ones? Lean meats don't have a thick layer of fat around the outside, or lots of white fatty streaks you can see. For ground meat (the kind you use for burgers or meat loaf), you need to read the label since the fat is mixed into the meat. Choose the packages

marked "lean" or "extra lean." For poultry, a lean choice is the white meat instead of the dark.

Vegetables and fruits especially offer a great deal of variety. To make choosing easier, eat by color! When you're at the produce department, choose fruits and vegetables of different colors: green, red, orange, and white. Eating by color means you're getting lots of different vitamins and minerals, and other good stuff.

Another way to think of food is: *Every Day Foods* and *Sometimes Foods*. Along with fruits and vegetables, Every Day Foods include cereals and grains; milk, and other low-fat dairy products; and meat, poultry, and fish. Then foods like sugar, fat, desserts, and soda are Sometimes Foods. Have these once in a while—for fun!

Be smart about how much you eat, and how often you eat it. *Balance* is what we call it. For instance, if you have a burger, fries, and some ice cream for lunch, that meal has a lot of fat. So, for dinner, try a low-fat meal, such as spaghetti and tomato sauce, a salad, some fruit, and graham crackers.

From time to time in this book, I'll have a Nutri Tip for you, giving you some further tips on nutrition. Minnie, too, will offer some helpful advice in her Minnie Minders, such as ideas for how to do things faster and easier in the kitchen. In fact, everyone wanted to do more than just give a recipe or two. They insisted on providing you with some advice on eating. Mostly, though, we all want you to have fun. Cooking is a terrific way to do that.

So, what are you waiting for? Let's get cookin'!

Mickey Mouse

5

Minnie's Minders

Hi, kids!

Ready to get started? Good. Before you begin there are a few things to know:

1 Read the recipe carefully. Make sure you have all the utensils and ingredients listed to make it.

2 Set everything you'll need out on a countertop or a table. Measure all the ingredients and put them on a large tray.

3 Reread the recipe to be sure you understand the steps. Double-check the utensils and ingredients.

4 Ask an adult to help if you have questions, or if there's something like draining hot pasta or slicing to be done. This sign **Ask an Adult** will appear next to steps where adult supervision is needed. Always ask if there's anything you're unsure about. Be safe, not sorry.

5 Tie back your hair if it's long. And, yes, all cooks wear aprons. Roll up your sleeves so they don't get dirty, caught on anything, or hang too close to a hot flame.

Remember: KEEP IT CLEAN. KEEP IT SAFE.

❶ Gather around the sink and wash your hands with warm, soapy water. (That goes for adult helpers, too!)

❷ Clean up spills and dribbles as you go along. That will help you avoid accidents and extra work later.

❸ Keep a damp sponge or some paper towels handy.

❹ Keep thick, dry oven mitts or pot holders nearby. They help prevent burns. Wet ones cause burns because the heat goes right through.

❺ Turn the handles of saucepans and skillets away from you and toward the middle of the stove when cooking. That keeps them from catching on something (or someone), or getting knocked over.

❻ When you've finished, put knives and other sharp objects aside and ask an adult to wash them separately. Then get to work washing and drying the other dishes and utensils, or load them into the dishwasher. Put away any utensils and foods you've used.

❼ Wash your hands again. Take off your apron and put it in the laundry.

❽ Remember to recycle. Separate the cans, foil, glass, and plastic containers from the other trash when you're cleaning up.

Throughout, I'll be offering hints and suggestions to make things easier for you.

Now you're ready to begin!

Minnie Mouse

Breakfast

❖ Belle's Favorite Eggs

❖ Donald's Muffin Breakfast Sandwich

❖ Daisy's Muffin Breakfast Sandwich

❖ Goofy's Smart-Start Oatmeal

❖ Three Caballeros' Breakfast Burrito

❖ Mickey's Whole Wheat Honey Pancakes

❖ Cinderella's Pumpkin Waffles

Belle's Favorite Eggs

Serves 1

UTENSILS

- ❖ Small saucepan
- ❖ Oven mitt
- ❖ Spoon
- ❖ Cutting knife
- ❖ Cutting board
- ❖ Small bowl
- ❖ Fork
- ❖ Kitchen scissors OR knife
- ❖ 1 tablespoon measuring spoon
- ❖ 1/2 teaspoon measuring spoon

INGREDIENTS

- ❖ 2 eggs
- ❖ 1/2 slice low-fat ham
- ❖ 1 tablespoon low-fat mayonnaise
- ❖ 1/2 teaspoon mustard

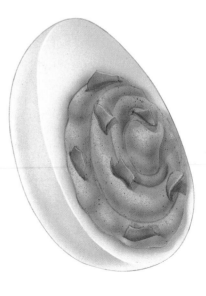

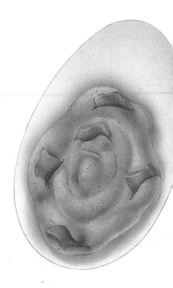

PLACE the eggs gently in saucepan. Cover with water, and bring to a boil over high heat. Turn heat to low, and simmer for 10 minutes.

(Ask an Adult) **USE** an oven mitt, and carefully take the pan off the heat. Place it in the sink, and pour out the water.

RUN cold water over the eggs in the pan. Let stand for 3 minutes.

CRACK the shells carefully with the back of a spoon. When the eggs are cool enough to handle, peel the eggs.

(Ask an Adult) **CUT** each of the eggs in half the long way to make 2 ovals.
Use a spoon to remove the yolks from the eggs.
Throw away one egg yolk. Place the other one in a small bowl.

(Ask an Adult) **CUT** the ham carefully into small pieces using the kitchen scissors OR knife. Place it in the bowl.

ADD the mayonnaise and mustard. Use a fork to mix them well.

SPOON the mixture into each of the egg halves.

Donald's Muffin Breakfast Sandwich

Serves 1

UTENSILS

❖ Toaster OR toaster oven

❖ Oven mitt

❖ Small plate

❖ 1 tablespoon measuring spoon

❖ 1 teaspoon measuring spoon

❖ Bread knife

INGREDIENTS

❖ 1 honey oat bran English muffin

❖ 1 tablespoon peanut butter

❖ 2 teaspoons honey OR jelly

❖ 1/2 small banana (peeled)

12

Mickey's Nutri Tip

Of course you can't eat cereal and milk in the car. But when I'm late, this is easy to wrap and eat along the way. Breakfast is my number one way to start the day.

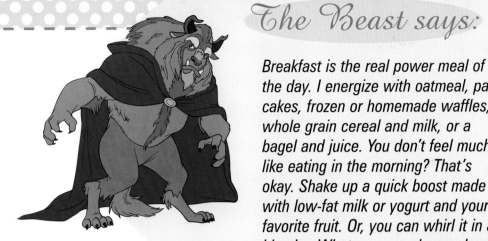

Breakfast is the real power meal of the day. I energize with oatmeal, pancakes, frozen or homemade waffles, whole grain cereal and milk, or a bagel and juice. You don't feel much like eating in the morning? That's okay. Shake up a quick boost made with low-fat milk or yogurt and your favorite fruit. Or, you can whirl it in a blender. Whatever you choose, breakfast is the heroic way to start the day.

SPLIT the English muffin gently in half. Toast until golden brown. Use an oven mitt and carefully remove it from the toaster.

PLACE half the muffin on a small plate. Spread the peanut butter on top.

SPREAD the honey OR jelly on the other half.

SLICE the banana carefully in half and place it on top of the honey OR jelly. Cover with the other half of the English muffin.

(Ask an Adult) **CUT** the sandwich in half, if desired.

13

Daisy's Muffin Breakfast Sandwich

Serves 1

UTENSILS

- ❖ Can opener (if using canned pineapple)
- ❖ Fork
- ❖ 1 paper towel folded in half 2 times to make a square
- ❖ Toaster OR toaster oven
- ❖ Oven mitt
- ❖ 1 tablespoon measuring spoon
- ❖ 1/4 teaspoon measuring spoon
- ❖ Bread knife

INGREDIENTS

- ❖ 1 whole pineapple ring (canned OR fresh*)
- ❖ 1 honey wheat English muffin
- ❖ 2 tablespoons low-fat cottage cheese OR low-fat cream cheese
- ❖ 1/4 teaspoon ground cinnamon
- ❖ 1/4 teaspoon sugar

*Look in the produce section of the supermarket for pineapples that are already peeled and cored. They're sold in a plastic bag.

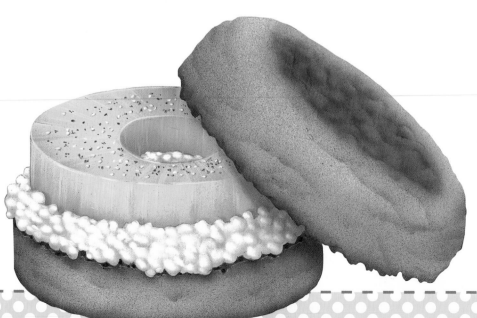

Be sure to check the setting on your toaster or toaster oven. Make sure it's on LIGHT. If the English muffin is too dark and crisp, it will be tricky to cut.

A bread knife has a serrated (bumpy) edge. That makes it easier to cut the bread without tearing it.

OPEN the can of pineapple rings if you don't have fresh pineapple. Use a fork to take out one of the rings and place it on a folded paper towel to drain.

SPLIT the English muffin gently in half, and toast lightly. Carefully remove with an oven mitt.

SPREAD the cottage cheese OR cream cheese on one half of the muffin. Place the pineapple ring on top.

PRINKLE the cinnamon and sugar evenly over the pineapple. Place the other half of the muffin on top.

 CUT the sandwich in half.

15

Mickey's Nutri Tip

Look at this! Cottage cheese in a sandwich. What a good way to get protein and calcium! For a super boost, have a glass of milk, too.

Goofy's Smart-Start Oatmeal

Serves 1

UTENSILS

- ❖ Small saucepan and cover
- ❖ 1 cup liquid measuring cup
- ❖ 1/4 cup dry measuring cup
- ❖ 1 tablespoon measuring spoon
- ❖ Spoon
- ❖ Oven mitt
- ❖ Small bowl
- ❖ 1 teaspoon measuring spoon

INGREDIENTS

- ❖ 1/2 cup water
- ❖ 1/4 cup old-fashioned oatmeal
- ❖ Pinch of salt
- ❖ 1/4 cup unsweetened chunky applesauce
- ❖ 2 tablespoons instant non-fat dry milk
- ❖ 1 tablespoon raisins
- ❖ Pinch of ground cinnamon
- ❖ 1 teaspoon sugar
- ❖ 1/3 cup low-fat milk

16

That fiber stuff everybody's talking about is important for kids, too. It's only found in plant and vegetable food. The funny part is fiber isn't really digested. But it helps move food through the body. Everybody should eat fruits, vegetables, whole grain cereals, and bread for fiber each day.

BRING the water to a boil in a small saucepan. Add the oatmeal and salt. Reduce the heat to low. Cook for 5 minutes, stirring occasionally.

 USE an oven mitt, and remove the pan from heat. Stir in the applesauce and dry milk. Add the raisins and cinnamon. Stir again and cover the pan. Let stand for 3 minutes.

SPOON the oatmeal into a small bowl to serve. Sprinkle with sugar, and pour the milk over the top.

Three Caballeros' Breakfast Burrito

Serves 1

UTENSILS

- ❖ Small microwave-safe bowl
- ❖ 1 tablespoon measuring spoon
- ❖ 2 forks
- ❖ Oven mitt
- ❖ 3 paper towels
- ❖ Microwave-safe plate OR paper plate

INGREDIENTS

- ❖ 1 egg
- ❖ 1 egg white
- ❖ 1 tablespoon milk OR water
- ❖ One 10-inch whole wheat OR flour tortill
- ❖ 2 tablespoons shredded low-fat cheese, any kind
- ❖ 2 tablespoons mild salsa

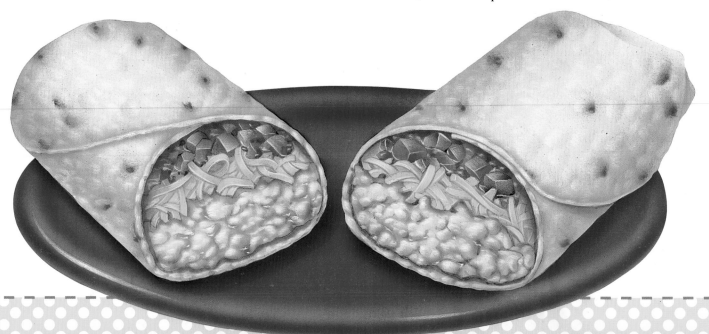

To separate an egg, place a small bowl on the kitchen counter. Gently crack the egg over the bowl, letting the white (liquid part) drop into the bowl. Keep the yolk (the yellow part) intact in one of the shell halves. Carefully slide the yolk from one shell half to the other allowing the white to drain into the bowl. This will take several attempts. Separating an egg takes some practice. You may need to do it a few times to get the knack.

COMBINE the egg, egg white, and milk OR water in a small microwave-safe bowl. Stir with a fork until blended.

ICROWAVE on HIGH for 30 seconds. Stir well with a fork. Microwave on HIGH again for 20 seconds. Stir again, and microwave once more for about 15 seconds, or until the eggs are cooked.

REMOVE the bowl from the microwave with an oven mitt. Cover the bowl with a paper towel.

PLACE the tortilla between two paper towels, and put it on a microwave-safe plate OR paper plate. Microwave on HIGH for 10 seconds. Use an oven mitt to remove from the microwave. Throw away the top paper towel.

SPOON the scrambled eggs onto the center of the flour tortilla. Use the side of a clean fork to spread the eggs, leaving a one-inch border on all sides. Sprinkle the cheese on top. Spoon the salsa on top of the cheese.

FOLD UP the bottom edge of the tortilla. Then fold the right edge over to just cover the eggs, and bring the left edge over that to make a burrito.

WRAP a napkin or paper towel around the bottom half, and eat the burrito as a sandwich.

Mickey's Whole Wheat Honey Pancakes

Makes twelve 3-inch pancakes

UTENSILS

- ❖ Medium mixing bowl
- ❖ Small mixing bowl
- ❖ 1/2 cup dry measuring cup
- ❖ 1/4 cup dry measuring cup
- ❖ 1/2 teaspoon measuring spoon
- ❖ 1/4 teaspoon measuring spoon
- ❖ 2 mixing spoons
- ❖ 1 cup liquid measuring cup
- ❖ 1 tablespoon measuring spoon
- ❖ Large skillet OR pancake griddle

INGREDIENTS

- ❖ 1/2 cup all-purpose flour
- ❖ 1/4 cup whole wheat flour
- ❖ 1/4 cup toasted wheat germ
- ❖ 1/2 teaspoon baking powder
- ❖ 1/2 teaspoon baking soda
- ❖ 1/4 teaspoon salt
- ❖ 1 egg
- ❖ 1 cup low-fat milk
- ❖ 2 tablespoons vegetable oil
- ❖ 2 tablespoons honey
- ❖ Nonstick cooking spray

COMBINE the flours, wheat germ, baking powder, baking soda, and salt in a medium bowl. Stir to mix it well.

COMBINE the egg, milk, oil, and honey in a small bowl. Stir to mix it well.

POUR the egg mixture into the flour mixture. Stir it gently until it is a smooth batter.

 HEAT a large skillet OR pancake griddle over medium-high heat. Spray it lightly with cooking oil.

 USE a 1/4 cup measuring cup—not completely full—to pour the batter onto the skillet or griddle. Cook the pancakes for about 3 minutes, or until little bubbles appear on top. Turn the pancakes over, and cook 1 to 2 minutes longer, until they're golden brown.

Mickey's Nutri Tip

What makes these pancakes special is the added wheat germ. It has a sweet, nutty taste. Wheat germ is a good source of vitamin E and fiber. Add it to cereal, ice cream, or yogurt for a crunchy treat.

Cinderella's Pumpkin Waffles

Makes four 7-inch waffles*

UTENSILS

- ❖ Waffle maker
- ❖ Large mixing bowl
- ❖ 1 cup dry measuring cup
- ❖ 1/2 cup dry measuring cup
- ❖ 1/4 cup dry measuring cup
- ❖ 1 teaspoon measuring spoon
- ❖ Small mixing bowl
- ❖ 1 cup liquid measuring cup
- ❖ 1 tablespoon measuring spoon
- ❖ 1 whisk OR 2 mixing spoons
- ❖ Soup spoon
- ❖ Fork
- ❖ Oven mitt

*Waffle makers come in different sizes. You may have to adjust this recipe to suit the one you have at home.

INGREDIENTS

- ❖ 1 1/2 cups all-purpose flour
- ❖ 1/4 cup packed dark brown sugar
- ❖ 2 teaspoons baking powder
- ❖ 1 teaspoon ground cinnamon
- ❖ 1 teaspoon salt
- ❖ 1 egg
- ❖ 1 1/4 cups low-fat milk
- ❖ 1/2 cup canned pumpkin (not pumpkin pie mix)
- ❖ 2 tablespoons vegetable oil
- ❖ Nonstick cooking spray

PREHEAT the waffle maker according to its directions.

COMBINE the flour, sugar, baking powder, cinnamon, and salt in a large bowl. Lightly whisk, or stir, to mix it well. You may need to crumble any bits of brown sugar with your fingers to be sure there are no lumps. Then whisk it again.

COMBINE the egg, milk, pumpkin, and oil in a small bowl. Whisk, or stir, to mix them well.

POUR the egg mixture into the flour mixture. Stir it gently until it forms a smooth batter.

SPRAY both sides of the waffle maker lightly with nonstick cooking spray.

USE a 1/2 cup measuring cup to pour a heaping amount of the batter onto the grids. Scrape out the inside of the cup with a soup spoon (this is a fairly thick batter). Use the spoon to spread the batter to the edges of the grids. Close the top, and cook for about 4 minutes (some waffle makers have a light that tells you when they're done), or until the waffle is set and the top opens easily. Use a fork to remove the waffle.

❖ Aladdin's Magic Carpet Rolls

❖ Chef Louis's Clam Chowder

❖ Goofy's Easy as A-B-C Soup

❖ Pumbaa's "Slimy Yet Satisfying" Chicken Noodle Soup

❖ Mickey's Neato Burrito

Aladdin's Magic Carpet Rolls

Serves 1

UTENSILS

❖ Cutting knife

❖ Cutting board

❖ Toothpicks

❖ Small dish

INGREDIENTS

❖ 2 slices soft whole wheat OR white bread

❖ 2 slices low-fat cooked ham

❖ 2 slices low-fat American OR Swiss cheese

❖ Mustard (for dipping)

PLACE the bread on a cutting board.

CUT *(Ask an Adult)* the crusts off both slices of bread.

PLACE 1 slice of ham and 1 slice of cheese on top of each slice of bread.

STARTING at one end, carefully roll up the bread, ham, and cheese, forming a long roll.

PUT a toothpick in the middle to hold it together.

MAKE another roll, using the other slice of bread, ham, and cheese.

SERVE the rolls with a small dish of mustard for dipping.

27

Mickey's Nutri Tip

Sometimes I use turkey or roast beef to make this. They're good protein foods, and the cheese gives me calcium. I have Peter Pan's Sparkling Pink Punch (page 72) to drink.

Chef Louis's Clam Chowder

Serves 4 (1 3/4 cups each)

UTENSILS

- ❖ Can opener
- ❖ Strainer
- ❖ Small bowl
- ❖ Cutting board
- ❖ Chopping knife
- ❖ Potato peeler
- ❖ 1 teaspoon measuring spoon
- ❖ 3-quart saucepan and cover
- ❖ 1 cup liquid measuring cup
- ❖ 1 large stirring spoon
- ❖ Ladle
- ❖ 4 small soup bowls

INGREDIENTS

- ❖ Two 6 1/2-ounce cans minced clams packed in clam juice
- ❖ 1 small onion
- ❖ 2 medium potatoes
- ❖ 1 teaspoon vegetable oil
- ❖ One 8-ounce bottle clam juice
- ❖ 3 cups low-fat milk
- ❖ Salt and pepper to taste

Be adventurous. Try new foods. I've tried Chinese noodles, Italian chicken, and different kinds of melon. Start with just one bite. Who knows? Maybe you'll discover a new favorite.

OPEN the clams and drain the juice into a small bowl. Set the clams and the juice aside.

(Ask an Adult) **PEEL** the onion, using a knife, and cut it into small pieces. (You should have about 1/2 cup.) Set aside.

(Ask an Adult) **USE** the peeler to peel the potatoes. Cut them into 1/2-inch cubes. Set aside.

(Ask an Adult) **HEAT** the oil in the saucepan over medium-low heat for 1 minute. Add the onion, and cook for 2 minutes, stirring frequently.

ADD the potatoes, the juice from the clams, and the bottle of clam juice. Stir well.

COVER the pot partially, and cook over medium heat for 15 minutes, stirring occasionally.

REDUCE the heat to low. Add the strained clams. Slowly pour in the milk, stirring constantly.

RE-COVER the pot partially, cook for 10 minutes longer. Add salt and pepper to taste.

(Ask an Adult) **LADLE** the soup carefully into bowls and serve.

29

Mickey's Nutri Tip

No doubt about it, I want to keep my bones strong and healthy. So I go for the calcium whenever I can. I like the way Chef Louis uses milk in this yummy soup.

Goofy's Easy as A-B-c Vegetable Soup

Serves 4 (1 1/2 cups each)

UTENSILS

❖ Can opener

❖ 3-quart saucepan and cover

❖ 1/4 teaspoon measuring spoon

❖ 1 cup dry measuring cup

❖ 1/2 cup dry measuring cup

❖ Kitchen scissors

❖ Ladle

❖ 4 small soup bowls

INGREDIENTS

❖ Two 13 3/4-ounce cans reduced-sodium chicken broth

❖ One 10-ounce package frozen mixed vegetables

❖ 1/4 teaspoon onion powder

❖ 1 cup canned drained cannellini beans (or your favorite type of canned beans)

❖ 1/2 cup alphabet macaroni (see Minnie Minder)

❖ 2 sprigs fresh parsley

❖ Grated Romano OR Parmesan cheese (optional)

Mickey's Nutri Tip

Good deal! Vegetables and beans are two of my favorite foods because they both have fiber. Beans are a bonus because they have protein, too. The macaroni is f-u-n, as Goofy would say.

30

OPEN the cans of chicken broth and pour into the saucepan. Add the vegetables and onion powder.

COVER and bring to a boil over high heat. Reduce the heat to medium. Cook for 5 minutes longer.

STIR in the beans and macaroni, and re-cover. Cook for about 10 minutes, or until the macaroni is tender, stirring occasionally.

TURN OFF the heat. Snip the parsley into small pieces with kitchen scissors and add to the soup.

 LADLE the soup into bowls. Sprinkle with grated cheese, if desired.

Minnie Minder

When I make this, I sometimes use other shapes of small macaroni. Little elbows or shells are nice. You can also find shapes like hearts, dinosaurs, and autumn leaves that come in colors, too!

31

Pumbaa's "Slimy Yet Satisfying" Chicken Noodle Soup

Serves 2 (about 1 1/2 cups each)

UTENSILS

- ❖ 2-quart saucepan
- ❖ 1 cup liquid measuring cup
- ❖ Can opener
- ❖ Medium bowl
- ❖ Fork
- ❖ Strainer
- ❖ Ladle
- ❖ 2 small soup bowls

INGREDIENTS

- ❖ 1 1/2 ounces bean thread noodles (see Minnie Minder)
- ❖ 2 cups warm water
- ❖ One 5-ounce can chunk white chicken, packed in water
- ❖ One 13 3/4-ounce can reduced-sodium chicken broth
- ❖ 3/4 cup water

PLACE the noodles in the saucepan, and cover with 2 cups *warm* water. Let soak (don't cook) for 10 minutes.

OPEN AND empty the can of chicken, and any juice, into a medium bowl. Use a fork to mash into small pieces. Set aside.

USE a fork to stir the noodles, when 10 minutes are up. Use a strainer to drain the noodles over the sink.

OPEN and pour the chicken broth and water into the saucepan. Bring to a boil over high heat.

ADD the noodles and the chicken. Reduce the heat to medium. Cook for 10 minutes longer.

Ask an Adult **LADLE** the soup carefully into bowls and serve.

You might need a fork to eat this soup. The bean-thread noodles are very slippery. That makes them tricky to eat. Look for them in the Oriental food section of your supermarket. They are also called Sai Fun, or cellophane noodles. Usually there are three "bundles" of noodles in a 3 3/4-ounce package. Use one of the bundles to make the soup. Angel hair pasta or cappellini can be used, too. Don't soak the pasta, just add it with the chicken.

Mickey's Neato Burrito

Serves 2 to 4

UTENSILS

- ❖ Fork
- ❖ Medium mixing bowl
- ❖ 1 cup dry measuring cup
- ❖ 1/2 cup dry measuring cup
- ❖ 1 microwave-safe plate
- ❖ Paper towel
- ❖ Oven mitt
- ❖ Small spatula

INGREDIENTS

- ❖ 3/4 cup low-fat refried beans
- ❖ 1/2 cup low-fat shredded cheddar cheese
- ❖ 1/2 cup mild salsa
- ❖ Four 6-inch whole wheat OR flour tortillas

Mickey's Nutri Tip

All you vegetarians will like this recipe. Beans, cheese, and tortillas (bread) are good foods to eat together.

USE a fork to combine the beans, cheese, and salsa in a medium bowl. Mix them well.

PLACE 1 tortilla on the counter. Loosely fill a 1/2 cup dry measuring cup with the bean mixture. Place mixture in the middle of the tortilla. Use a fork to form a line about 2 inches wide down the center. Leave about 1 inch at the top and bottom.

FOLD UP about 1 inch on the bottom of the tortilla to keep the filling in. Bring the folded bottom edge up to the center and then bring the top edge down to close. Tuck the open ends under.

PLACE the tortilla roll on the plate with the tucked sides down.

REPEAT with the other 3 tortillas. Place them on the plate so they look like the spokes of a wheel.

MICROWAVE on HIGH for 45 seconds. Use oven mitt to remove the plate from the microwave. Let stand for 1 minute. Use a small spatula to serve the burritos.

35

Minnie Minder

It's very important to wait a minute (or two) when you take food out of a microwave. Because it cooks so fast it needs time to finish getting done. Be careful when you take a bite. The food can be hotter than you think.

Smart microwave cooks use a few seconds less—not more—when cooking to avoid burns, overcooking, and possibly spoiling a recipe.

Snacks & Side Dishes

- ❖ The Sultan's Hummus Dip with Veggies

- ❖ Abu's Baked Apples

- ❖ Rafiki's Coconut Fruit Kabobs

- ❖ Dumbo's Broccoli Slaw

- ❖ Thumper's Buttered Baby Carrots

- ❖ Alice's Cheesy Rice

- ❖ Pocahontas's Triple Corn Bread

The Sultan's Hummus Dip
with Fresh Vegetables

Serves 3 to 4 (Makes 1 1/2 cups)

UTENSILS

- ❖ Can opener
- ❖ Strainer
- ❖ Food processor OR blender
- ❖ 1 tablespoon measuring spoon
- ❖ 1/4 teaspoon measuring spoon
- ❖ Rubber scraper OR spoon
- ❖ Small bowl
- ❖ Medium plate

INGREDIENTS

- ❖ One 15-ounce can chickpeas (garbanzo beans)
- ❖ Juice of 1/2 lemon (about 2 tablespoons)
- ❖ 1/4 teaspoon garlic powder
- ❖ 3 sprigs fresh parsley
- ❖ 1/4 cup low-fat plain yogurt

FOR DIPPING

An assortment of any fresh vegetables: carrot sticks, celery sticks, green or red pepper slices, cucumber slices; or triangles of whole wheat pita bread

Mickey's Nutri Tip

This is so good I sometimes have it for lunch or dinner. You can, too. Just spoon some of the hummus into a pita pocket. Then put in some veggies. Have it with a glass of milk. Wow! A complete meal in a pocket!

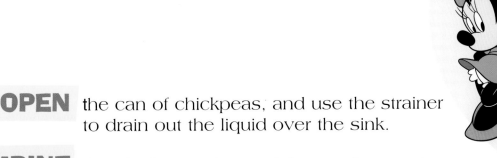

Here's a chance to go to the grocery salad bar. Pick your favorite vegetables. Everything is already washed and sliced for you.

Remember to play it safe when using a food processor or blender. Handle the blades carefully, or ask for help. Make sure the lid is in place before turning on the motor. Always unplug the machine before opening the lid or removing the container from the base.

OPEN the can of chickpeas, and use the strainer to drain out the liquid over the sink.

COMBINE the chickpeas, lemon juice, garlic powder, parsley, and yogurt in the workbowl of a food processor (or the container of a blender). Lock the top, or put the lid firmly in place.

PUREE until smooth. Once or twice, turn off the machine and scrape the sides of the workbowl or container.

SPOON the hummus mixture into a small bowl.

PLACE the fresh vegetables, or pita triangles, on a plate. Dip into the hummus and enjoy!

39

Abu's Baked Apples

Serves 2

UTENSILS

- Plastic apple corer (see Minnie Minder)
- Small microwave-safe baking dish
- 1/4 cup dry measuring cup
- 1/2 teaspoon measuring spoon
- Plastic wrap
- Oven mitt
- Serving spoon

INGREDIENTS

- 2 large apples
- 1/4 cup apple juice OR apple cider
- 1/4 cup maple syrup
- 1/2 teaspoon ground cinnamon

40

Bambi says:

Think you get vitamin C just from eating oranges or drinking juice? Well, look at the lineup. Tomatoes, green peppers, pineapples, peaches, cantaloupes, and tangerines are some of the vitamin-C-rich foods to try. Kiwi, raspberries, strawberries, spinach, and potatoes are other good choices, too. Have at least one serving every day.

USE the apple corer to remove the core from each apple.

PLACE the apples in a small microwave-safe dish.

POUR the apple juice and maple syrup over the apples and sprinkle the cinnamon on top.

COVER the baking dish loosely with plastic wrap.

PLACE the baking dish in the microwave. Cook the apples on HIGH for about 4 minutes. (If your microwave doesn't have a rotating bottom, turn the dish a 1/2 turn after 2 minutes. Then cook for 2 minutes longer.)

USE oven mitt to remove the dish from the microwave. Let stand (covered) for 3 minutes before serving. Carefully remove the plastic wrap.

PLACE an apple in a small dish, and spoon some of the cooking syrup over the top.

Minnie Minder

You can buy a plastic apple corer that is easy to use. It's safer than a knife.

When removing plastic wrap from cooked foods, start from the side of the dish furthest from you and peel it back. Turn your head to the side in case there is still any steam under the wrap. If you peel the apples before cooking, reduce the time to 2 1/2 minutes.

41

Mickey's Nutri Tip

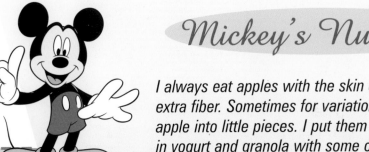

I always eat apples with the skin on. That way I get extra fiber. Sometimes for variation I cut the cooked apple into little pieces. I put them in a bowl and stir in yogurt and granola with some of the cooking syrup mixed in.

Rafiki's Coconut Fruit Kabobs

Serves 2

UTENSILS

- ❖ Can opener
- ❖ Medium bowl
- ❖ Chopping knife
- ❖ Cutting board
- ❖ Plastic apple corer
  (See page 43)
- ❖ Large spoon
- ❖ 4 small wooden skewers
  (about 6 inches long)
- ❖ 1 medium plate
- ❖ 1/4 cup dry measuring cup

INGREDIENTS

- ❖ One 8-ounce can pineapple chunks, packed in juice
- ❖ 1 medium banana, peeled
- ❖ 1 medium apple
- ❖ 1/4 cup orange juice
- ❖ 1/4 cup shredded coconut

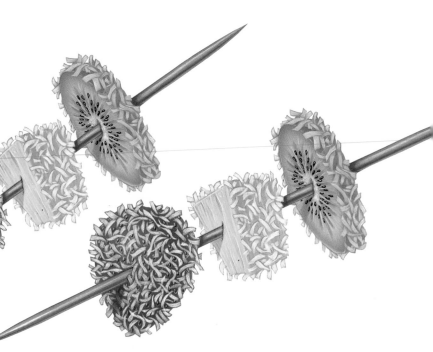

You can cover these kabobs with plastic wrap and put them in the refrigerator. Then your snack is ready when you are. Eat these kabobs within 24 hours.

OPEN the can of pineapple chunks. Pour the pineapple chunks and juice into a medium bowl.

 SLICE the banana into 1-inch pieces. Add to the bowl.

 **USE** apple corer to remove the core from the apple. Cut the apple into 1-inch cubes. Add to the bowl.

POUR the orange juice over the fruit and stir to blend.

PUT the fruit on a skewer, alternating the chunks of pineapple, banana, and apple. Fill up each skewer with an equal amount of fruit.

POUR half the coconut onto a plate. Place the skewers on top. Sprinkle remaining coconut over the fruit. Gently press the coconut to coat the fruit on all sides.

43

OTHER FRUIT POSSIBILITIES

cantaloupe	grapes	papaya
strawberries	nectarines	watermelon
pears	kiwi	peaches

Dumbo's Broccoli Slaw

Serves 4

UTENSILS

❖ Medium bowl

❖ 1 cup dry measuring cup

❖ 1 cup liquid measuring cup

❖ Mixing spoon

INGREDIENTS

❖ 2 cups broccoli slaw OR cabbage cole slaw (see Minnie Minder)

❖ 1/4 cup seasoned rice vinegar OR white vinegar with 1 tablespoon sugar added

❖ Salt and pepper to taste

44

Baloo says:

In the jungle we eat lots of fruits and veggies. They're colorful and delicious. Eat at least five servings of these crispy-crunchy, sweet-juicy treats every day. Need some ideas? Here are a few:

Whirl your favorite fruit (strawberries, peaches, kiwi) with low-fat yogurt or milk in a blender.

Stir fresh or frozen veggies—of all different colors—into rice, pasta, and soups at the end of the recipe. Cook a few minutes longer to be sure they're heated through.

Visit the salad bar in restaurants and supermarkets. Pick lots of different color vegetables to get the best variety.

Top pancakes (Mickey's favorite), waffles, toast, bagels, yogurt, or whole grain cereals with chopped or dried fruits.

Toss a can (or box) of 100% pure juice or a bag of cut up veggies or dried fruits into your backpack or gym bag for an easy snack.

PLACE the broccoli slaw in a medium bowl.

ADD the vinegar. Mix well.

SEASON with salt and pepper to taste.

45

Thumper's Buttered Baby Carrots

Serves 4

UTENSILS

- ❖ 1-quart saucepan and cover
- ❖ 1 cup liquid measuring cup
- ❖ 1 knife
- ❖ 1 fork
- ❖ Kitchen scissors
- ❖ Toothpicks, if desired

INGREDIENTS

- ❖ One 16-ounce package peeled, ready-to-eat baby carrots
- ❖ 1/4 cup orange juice
- ❖ 1/4 cup water
- ❖ 2 tablespoons salted butter OR margarine
- ❖ 2 sprigs fresh dill OR parsley
- ❖ Salt to taste

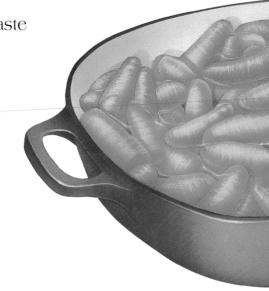

Mickey's Nutri Tip

I bet everyone always tells you that carrots are good for your eyes. Guess what? They're right. It's because carrots have lots of vitamin A. Vitamin A keeps muscles strong, too. It even keeps gums and teeth healthy.

COMBINE the carrots, juice, water, and butter in a 1-quart saucepan.

 COVER the pan, and cook over medium-low heat for 10 to 15 minutes or until the carrots are tender. (Use a fork to test when they're done.)

UNCOVER the pan when the carrots are tender, and cook a few minutes longer, until most of the juice is gone.

 **USE** the kitchen scissors to snip the dill, or parsley, into the pan. Sprinkle with salt, if desired. Stir and serve. (It's fun to eat these carrots with toothpicks.)

Minnie Minder

Look on the wrapper of the butter or margarine. You will see lines that show how much equals a tablespoon. Use a knife and press lightly on the paper at 2 table-spoons. Open the wrapper and cut the piece you need.

47

Mushu says:

Snacks are smart. Between-meal treats are a good way to keep your energy going. Snacks fuel up your body like gas fuels up a car. To keep your engine running smoothly try some of these:

- *Rice crackers with PB & J*
- *Whole grain cereal (not just for breakfast!) with milk and fresh fruit or berries*
- *Giant fresh strawberries drizzled with a little chocolate syrup*
- *Air-popped popcorn sprinkled with parmesan cheese*
- *Banana slices covered with peanut butter and chopped dry-roasted peanuts*
- *Whole wheat pita triangles (one round bread cut in four pieces) topped with low-fat whipped cream cheese or peanut butter, or dunked in your favorite salsa or dip*
- *Frozen grapes (after you wash them, pat dry with a paper towel or clean kitchen towel. Remove the stems, then place the grapes in a single layer in a shallow pan, and place in the freezer.) A simple frosty treat!*

Alice's Cheesy Rice

Serves 4

UTENSILS

- ❖ 1 1/2-quart saucepan
- ❖ 1 cup liquid measuring cup
- ❖ 1 cup dry measuring cup
- ❖ Mixing spoon
- ❖ 1/2 cup dry measuring cup
- ❖ Oven mitts

INGREDIENTS

- ❖ 2 cups water
- ❖ 1 cup long-grain rice
- ❖ 3/4 cup shredded cheddar OR Monterey Jack cheese
- ❖ 1/2 cup reduced-fat sour cream

Mickey's Nutri Tip

Alice loves cheese because it tastes so good, and rice gives her lots of energy. When I make this I like to add some (thawed) frozen peas or cooked carrots when I stir in the cheese and sour cream. That makes it look pretty, and we get vegetables at the same time!

BRING the water to a boil over high heat in a 1 1/2-quart saucepan.

STIR in the rice. Reduce heat to low. Cover and cook for about 20 minutes, or until most of the liquid is absorbed. Ask for help when checking the rice. Be sure to use oven mitts when taking the cover off the pot, and keep your face away when opening it (see Minnie Minder, page 41).

ADD the cheese and sour cream to the rice. Stir well until it is completely blended. Cover, and let stand for 1 minute so the cheese melts completely.

Minnie Minder

There are more and more preshredded cheeses in the dairy case every day. Try them all. You can also grate your favorite cheese by hand. Be careful. Scraped knuckles are not part of the recipe.

49

Nala says:

Protein makes me strong. You can be, too. Eat 2 to 3 servings each day. Just one chicken leg, a small burger, 1 egg, 1/2 cup cooked beans, or 1/3 cup of nuts is a serving. Make lean—less fat—choices for trim bodies.

Pocahontas's Triple Corn Bread

Makes one 9-inch square pan

UTENSILS

- ❖ One 9-inch square pan
- ❖ Strainer
- ❖ Small bowl
- ❖ Large mixing bowl
- ❖ Medium mixing spoon OR whisk
- ❖ Medium mixing bowl
- ❖ 1 cup dry measuring cup
- ❖ 1 teaspoon measuring spoon
- ❖ 1/2 teaspoon measuring spoon
- ❖ 1 cup liquid measuring cup
- ❖ 1 tablespoon measuring spoon
- ❖ Oven mitt
- ❖ Wire baking rack
- ❖ Cutting knife

INGREDIENTS

- ❖ Nonstick cooking spray
- ❖ One 7-ounce can Mexicorn niblets
- ❖ 1 cup yellow stone-ground cornmeal
- ❖ 1 cup all-purpose flour
- ❖ 2 teaspoons baking powder
- ❖ 1/2 teaspoon baking soda
- ❖ 1/2 teaspoon salt
- ❖ 1 egg
- ❖ 1 cup low-fat buttermilk OR plain yogurt
- ❖ 1/3 cup corn oil
- ❖ 2 tablespoons honey

PREHEAT oven to 425°F
(400°F if using a glass pan).

SPRAY the baking pan with nonstick
cooking spray. Set aside.

USE a strainer, with a small bowl underneath to
catch the liquid, to drain the corn. Set
aside the corn. Throw away the liquid.

COMBINE the cornmeal, flour, baking powder, baking soda,
and salt in a large bowl. Stir (or whisk) to mix it well.

USE the same spoon or whisk to combine the egg, buttermilk (OR
plain yogurt), oil, and honey in a medium bowl until it is blended.

POUR the egg mixture into the cornmeal mixture and stir just until it is
combined. (This is a dry, thick batter.) Stir in the corn. (Don't
overmix this batter. It needs just enough stirring to mix all the
ingredients together.)

SCRAPE the mixture into the prepared pan.
Lightly smooth the top to make an even layer.

BAKE for 20 minutes, or until golden brown, and the bread pulls away
from the sides of the pan.

REMOVE the pan from the oven using oven mitt. Place the pan on a wire
rack to cool for about 10 minutes. Carefully cut the bread into
squares. This bread is very good served warm.

Minnie Minder

You can make a substitute for buttermilk or yogurt if you don't have any. In a 1 cup liquid measuring cup combine 1 tablespoon of lemon juice and just enough milk to make 1 cup. Let that stand for 10 minutes, then use as directed.

When measuring the honey, rub a little oil on the spoon first. Then the honey will slide right off the spoon, and into the bowl.

51

Dinner

❖ Lady's Favorite Pasta

❖ Tramp's Favorite Pasta

❖ Mulan's Fresh Fish Sticks

❖ Hercules' Powerhouse Meat Loaf

❖ Pluto's Power-Packed Pizza

❖ Mufasa's Mouthwatering Chops

Lady's Favorite Pasta

Serves 6 to 8

UTENSILS

- ❖ Small cutting knife
- ❖ Cutting board
- ❖ Can opener
- ❖ Medium microwave-safe bowl
- ❖ Plastic wrap
- ❖ Oven mitt
- ❖ 6-quart pot
- ❖ Large strainer OR colander
- ❖ Slotted spoon (a cooking spoon with holes)

INGREDIENTS

- ❖ 1 large garlic clove
- ❖ One 13 3/4-ounce can reduced-sodium chicken broth
- ❖ 2 tablespoons olive oil
- ❖ One 16-ounce package dried cappellini pasta
- ❖ 1 cup grated Parmesan cheese
- ❖ 1/2 cup fresh basil leaves OR parsley leaves

54

PEEL the garlic, and cut it into 4 pieces.

OPEN the can of chicken broth and pour it into a medium microwave-safe bowl. Cover the bowl with plastic wrap. Microwave on HIGH for 3 minutes.

REMOVE the bowl from the microwave using the oven mitt. Carefully remove the plastic wrap (see Minnie Minder, page 41). Add the olive oil to the broth, and set aside.

FILL a 6-quart pot with about 4 quarts of water. Bring to a boil over high heat.

ADD the pasta carefully. Stir. Cook the pasta for 10 to 12 minutes until tender, or according to the package directions.

DRAIN the pasta over the sink using a large strainer, or colander. Then return the pasta to the pot.

REMOVE the garlic pieces from the broth mixture using the slotted spoon. Carefully pour the broth mixture, and half of the Parmesan cheese, over the pasta in the pot. Stir it well.

SERVE with the remaining Parmesan cheese sprinkled on top. You can also place a few fresh basil leaves OR parsley sprigs over each dish to make it look attractive.

Minnie Minder

It's easy to forget that water can be heavy—especially big pots of it. Always ask for help when cooking pasta.

The piece of garlic that you buy in the store is called a bulb, or head, of garlic. Each one of the sections that can be separated is called a clove.

Pumbaa says:

Rice, noodles, tortillas, bread, oatmeal, spaghetti, couscous, pretzels, and cold cereal are the foods to eat the most of each day. Just one slice of bread, a half cup of cooked rice or macaroni, one tortilla, or 3/4 cup of cold cereal is a serving from this group. Add 'em up. Six isn't so much after all.

Tramp's Favorite Pasta

Serves 6 to 8

UTENSILS

- ❖ 6-quart pot
- ❖ 1-quart saucepan
- ❖ Can opener
- ❖ Large cooking spoon
- ❖ Large strainer OR colander

INGREDIENTS

- ❖ One 24-ounce jar mild salsa
- ❖ One 15-ounce can black beans, drained
- ❖ One 16-ounce package multicolored fusilli pasta
- ❖ 1 cup shredded low-fat Monterey Jack cheese OR shredded low-fat cheddar cheese
- ❖ 1 cup fresh cilantro OR coriander leaves
- ❖ 1 cup fat-free sour cream (if desired)

Mickey's Nutri Tip

Don't forget that beans are a great source of protein. Packed with fiber, they also count as a vegetable and come in lots of different colors and sizes. What a bonus!

Look in the dairy section for cheese that is already shredded to make this recipe easier. You can grate your own, but be careful of fingers and knuckles. No scrapes!

 BRING 4 quarts of water to a boil in a 6-quart pot.

POUR the jar of salsa into a 1-quart saucepan. Add the beans. Simmer on low heat for about 10 minutes, stirring occasionally.

 ADD the pasta carefully to the boiling water. Stir. Cook the pasta for 12 to 15 minutes, until tender, or according to package directions.

 DRAIN the pasta over the sink using a large strainer, or colander. Then return the pasta to the pot.

 ADD the salsa and beans to the pasta. Stir carefully.

SERVE the pasta with shredded cheese and cilantro OR coriander leaves sprinkled on top. (You might like to have some sour cream on the side, for dipping the pasta.)

57

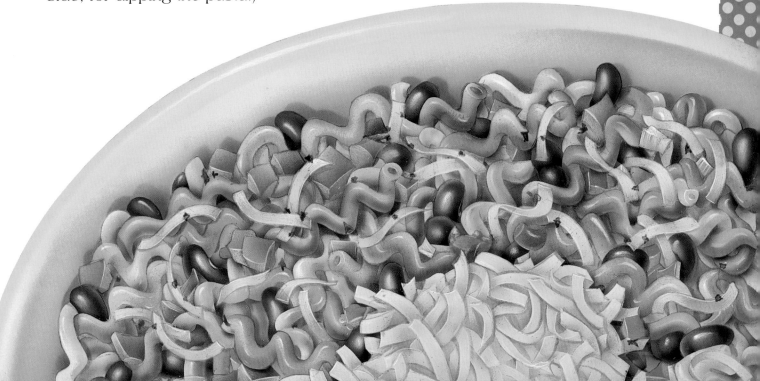

Mulan's Fresh Fish Sticks

Serves 4

UTENSILS

❖ Baking sheet

❖ Cutting board

❖ Kitchen scissors OR knife

❖ Medium bowl

❖ 1 tablespoon measuring spoon

❖ Plastic food storage bag

❖ Oven mitt

INGREDIENTS

❖ Nonstick cooking spray

❖ 1 pound flounder, catfish, OR white fish fillets

❖ 2 egg whites

❖ 2 tablespoons water

❖ 1 cup seasoned bread crumbs

❖ Tartar sauce (if desired)

PREHEAT the oven to 400°F. Spray a baking sheet with cooking spray. Set aside.

PLACE the fish fillets on a cutting board. Carefully cut them into one-and-a-half-inch strips using kitchen scissors OR knife.

MIX the egg whites and water together in a medium bowl until blended.

POUR the bread crumbs into a plastic bag.

DIP each fish strip into the egg mixture. Then drop each piece into the plastic bag. Close the bag, or seal it with a tie. Shake the bag to coat all the fish strips with bread crumbs.

PLACE each coated strip on the baking sheet. Leave some space in between each one.

59

BAKE for about 15 minutes, or until slightly browned and crisp.

REMOVE the baking sheet from the oven, using oven mitt.

SERVE with tartar sauce, if desired.

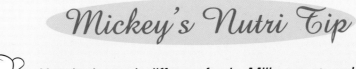

Mickey's Nutri Tip

Your body needs different foods. Milk, yogurt, and cheese build strong bones. Fish, lean pork, chicken, and beans have protein for muscles. Carrots, broccoli, and sweet potatoes keep your eyes, skin, and hair healthy. For lasting energy eat rice, whole wheat English muffins, and whole grain cereals. In other words, eat a variety of foods every day.

Hercules' Powerhouse Meat Loaf

Serves 4 to 6

UTENSILS

- Potato peeler
- Cutting board
- Chopping knife
- Food processor
- 1 cup measuring cup
- 1/4 teaspoon measuring spoon
- 1/8 teaspoon measuring spoon
- 1 soup spoon
- Nonstick 12-muffin-cup pan OR 2 six-muffin-cup pans
- Oven mitt
- Meat thermometer
- 2 forks

INGREDIENTS

- 1 medium potato
- 1 small onion
- 1 medium tomato
- 1 stalk celery
- 10 peeled ready-to-eat baby carrots (OR 2 large carrots, washed and peeled)
- 1/2 pound lean ground beef
- 1 cup packaged bread crumbs
- 2 egg whites
- 1/4 teaspoon salt
- 1/8 teaspoon pepper
- Mustard and ketchup in squeezable bottle:

If your muffin pan is not a nonstick kind, spray each little cup with nonstick cooking spray first. That helps each meat loaf come out of the pan easily.

A meat thermometer is the BEST way to be sure meat is cooked right. That's important because sometimes bacteria in raw meat can make you sick.

PREHEAT oven to 325°F.

PEEL the potato, using a potato peeler. Set aside.

CUT a thin slice from each end of the onion. Take off the outside skin. Set aside.

CHOP each of the vegetables (potato, onion, tomato, and celery) on the cutting board into 4 or 5 pieces.

PUT all the vegetables (including the baby carrots) into the workbowl of a food processor.

ADD the meat, bread crumbs, egg whites, salt, and pepper.

PLACE the lid on the food processor, and lock it in place. Turn on the processor. Let it run for about 1 to 2 minutes, until all the vegetables are chopped and combined with the meat.

TURN OFF the food processor. Unplug the cord, and remove the lid.

FILL each of the muffin cups with the meat loaf mixture using the soup spoon (see Minnie Minder).

DECORATE the top of each one with mustard and ketchup.

PLACE the pan carefully in the oven using oven mitt. Bake for 45 minutes, or until a meat thermometer placed in the center reads 160°F.

REMOVE the muffin pan carefully from the oven, using oven mitt. Gently remove each meat loaf using one fork underneath to lift it out, and another fork on top to hold it steady.

61

Pluto's Power-Packed Pizza

Serves 4

UTENSILS

- ❖ Cutting board
- ❖ Slicing knife
- ❖ 2 nonstick baking sheets
- ❖ 1 cup measuring cup
- ❖ 1 tablespoon measuring spoon
- ❖ Large bowl
- ❖ 1 pizza cutter OR cutting knife
- ❖ Oven mitts

INGREDIENTS

- ❖ One 10-ounce package refrigerated pizza dough
- ❖ 1 medium tomato
- ❖ 1 small onion
- ❖ 1/2 cup prepared tomato sauce OR mild salsa
- ❖ 1 1/2 cups lettuce, torn into small pieces
- ❖ 8 small pitted black olives
- ❖ 2 ounces shredded pizza cheese (1/2 cup)
- ❖ 2 tablespoons prepared fat-free Italian dressing

FOR THE PIZZA CRUSTS

Minnie Minder

Surprise! No melted cheese on this pizza. It's fun because you can toss everything together in a big bowl, and plop it on. Or, make layers, and sprinkle the dressing on top.

Buy cheese that is already shredded. The pizza cheese is mozzarella and provolone mixed together. Other cheeses, like cheddar, mozzarella, and Monterey Jack, come shredded, too. Another trick is to go to a grocery salad bar to pick the vegetables for this pizza. They're all ready for you to use. You can try new ones each time.

Get ready for a messy meal! You may need a few extra napkins for help.

PREHEAT the oven to 425°F.

OPEN the package of pizza crust. Gently remove the dough. Do not unroll.

PLACE the dough on the cutting board. Use a knife to cut the dough into 4 equal pieces.

PLACE 1 piece on a baking sheet. Press it down with the palm of your hand. Turn the dough over and press it again.

USE your fingers to gently, yet firmly, press the dough into a 6-inch circle.

REPEAT with the other 3 pieces. (2 crusts on each sheet)

PLACE both baking sheets carefully in the oven. Bake for about 7 minutes, or until the crusts are lightly browned.

REMOVE the baking sheets from the oven carefully, using oven mitts.

LET COOL on the baking sheet for about 10 minutes, or until they are not too hot to touch. (If any of the crusts are puffed up press down lightly after they cool.)

(continued on next page)

63

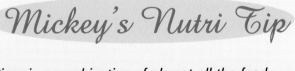

Mickey's Nutri Tip

Pizza is a combination of almost all the food groups! You could also add sliced ham, or chicken, too. That would put in the protein, and give you all five groups.

(continued from previous page)

FOR THE PIZZA TOPPINGS

METHOD 1

PLACE the baked pizza shells on a cutting board or a kitchen counter. Place 2 tablespoons of the tomato sauce OR salsa over the top of each one. Use the back of the spoon to make a smooth, even layer of sauce. Set aside.

 PEEL the onion. Cut it into thin slices or chop it into small pieces. Put it into a large bowl.

 CUT the tomato into small pieces. Put it into the bowl.

ADD the rest of the ingredients to the bowl. Use your (clean!) hands to toss well.

PLACE 1/4 of the mixture on top of each pizza.

USE both your hands, and carefully place the finished pizza on a medium-size plate. Forget the knife and fork on this one. Have fun.

METHOD 2

PLACE the baked pizza shells on a cutting board or a kitchen counter. Place 2 tablespoons of the tomato sauce OR salsa over the top of each one. Use the back of the spoon to make a smooth, even layer of sauce. Set aside.

 PEEL the onion. Cut it into thin slices or chop it into small pieces. Set aside.

 CUT the tomato into thin slices or chop it into small pieces. Set aside.

PLACE a layer of each of the vegetables and the cheese evenly on each pizza. Sprinkle 1 tablespoon of the dressing over the top.

USE both your hands and carefully place the finished pizza on a medium-size plate. Forget the knife and fork on this one. Have fun.

Pongo says:

Know Your Nutrients! Just six simple things—nutrients—are found in foods. They're very important for growing up strong and healthy.

- Protein comes from lean pork, chicken, turkey, beef, beans, nuts, and eggs. Proteins help build, maintain, and repair the body.
- Carbohydrates fuel us up. Eat rice, crackers, tortillas, whole grain cereals, and breads, along with fruits and vegetables for lots of energy all day long.
- Fats provide lots of energy, too, and everyone's body needs a little. Use high-fat foods like mayonnaise, salad dressing, and whipped cream sparingly.
- Vitamins and minerals help get lots of things done for your body. They make strong bones and teeth, shiny hair, and smooth skin. Vitamins are named after letters of the alphabet, like A, C, D, and E. Calcium, iron, and potassium are some of the important minerals.
- Water is a nutrient, too. In fact, it's the most important one of all. And most of us don't drink enough water. Have at least four glasses each day. Try some when you're really feeling tired and see how good you feel. And, of course, *always* stop and take a sip whenever you see a water fountain.

65

Mufasa's Mouthwatering Chops

Serves 4

UTENSILS

- ❖ 1 large (12-inch) nonstick frying pan and cover
- ❖ 1 teaspoon measuring spoon
- ❖ Fork
- ❖ 1 cup liquid measuring cup
- ❖ Spatula

INGREDIENTS

- ❖ 4 boneless pork chops, each about 1-inch thick
- ❖ 1 teaspoon dried Italian seasoning
- ❖ 1/2 cup reduced-sodium chicken broth
- ❖ 1 tablespoon lemon juice
- ❖ Topper (see Mickey's Nutri Tip)

66

Mickey's Nutri Tip

Boneless pork chops have hardly any fat, and they're easy to eat. Here's how some of my pals like to top their chops. Have you got more ideas?

Jose's Mexican Fiesta
Mild salsa and grated Monterey Jack cheese.

Baloo's Bear Necessities
Canned (drained) pineapple slices with brown sugar.

Wendy's Honey of a Peach
Canned (drained) sliced peaches drizzled with honey.

Pinocchio's Honest-to-Goodness Favorite
Pizza sauce sprinkled with grated cheese.

The Beast's Feast
Mashed potatoes with grated cheddar cheese.

Pork is tender and juicy when you don't overcook it. These chops cook fast because there's no bone, and hardly any fat.

HEAT the frying pan over medium-high heat for 1 minute.

(Ask an Adult) **PLACE** the chops in the pan, carefully.

COOK for 2 minutes. Use a fork to turn the chops over. Cook 2 minutes longer.

SPRINKLE the Italian seasoning over the chops. Pour in the chicken broth, and lemon juice.

(Ask an Adult) **ADD** your favorite topper.

REDUCE heat to low. Cover, and cook for 2 minutes.

(Ask an Adult) **TRANSFER** the chops, using the spatula, to serving dishes.

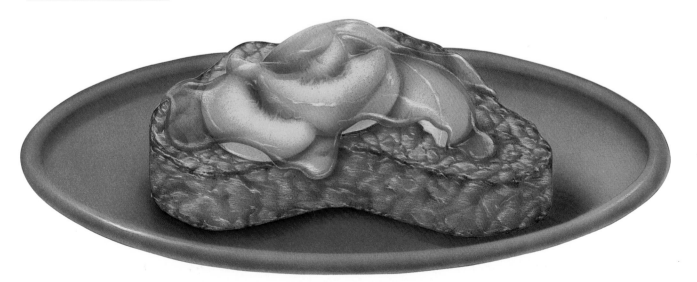

Beverages

❖ Mrs. Potts's Spiced Cider

❖ Peter Pan's Sparkling Pink Punch

❖ Jasmine's Fruitie Smoothie

Mrs. Potts's Spiced Cider

Serves 2 (about 1 1/4 cups each)

UTENSILS

- ❖ 1 cup liquid measuring cup
- ❖ Small saucepan
- ❖ Oven mitt
- ❖ Slotted spoon
- ❖ 2 mugs

INGREDIENTS

- ❖ 1/2 cup orange juice
- ❖ 1 cinnamon stick, about 2-inches long
- ❖ 2 whole cloves
- ❖ 2 cups apple cider OR apple juice

COMBINE the orange juice, cinnamon stick, and cloves in the saucepan.

 BRING to a boil over high heat. Reduce heat to low, and simmer for 10 minutes.

 ADD the cider (or juice). Cook over high heat for 1 minute. Using an oven mitt, carefully remove the pan from the heat.

REMOVE the cinnamon stick and cloves using a slotted spoon.

 POUR the cider carefully into two mugs.

Mickey's Nutri Tip

When it's chilly outside, this keeps me warm inside. Mrs. P's juice combo is a vitamin C special.

Peter Pan's Sparkling Pink Punch

Serves 6 (about 1 cup each)

UTENSILS

- ❖ Can opener
- ❖ Large pitcher OR punch bowl and ladle
- ❖ Mixing spoon
- ❖ Plastic wrap
- ❖ Punch cups OR glasses

INGREDIENTS

- ❖ 2 1/2 cups unsweetened pineapple juice
- ❖ 2 cups pink grapefruit juice
- ❖ 1 1/2 cups water or sparkling water, chilled
- ❖ Grenadine syrup to taste (if desired)
- ❖ Fruit slices for garnish: orange slices, pineapple slices, kiwi slices, or whole strawberries (if desired)

Minnie Minder

Grenadine syrup is a deep red syrup that's used to color and flavor drinks. It is very sweet. Add a tablespoon, then taste the punch. If you need more, add another teaspoon.

This punch is not too fizzy. You can use water instead of the sparkling water if you want a bubble-free punch.

OPEN the cans of pineapple and grapefruit juice.

POUR the pineapple and grapefruit juices into a large pitcher. Stir.

COVER with plastic wrap and refrigerate until chilled.

STIR in the water OR sparkling water and some grenadine, if desired. Pour, or ladle, the punch into glasses.

GARNISH with a slice of fruit, if desired.

100% pure is what I look for when choosing juice. That means no added sugar or water. I like my vitamins and minerals straight from the source. You can use your favorite blend of juices. Just make sure they're pure.

Jasmine's Fruitie Smoothie

Serves 4 (about 1 cup each)

UTENSILS

❖ Blender OR food processor

❖ 1 cup liquid measuring cup

❖ 1/4 teaspoon measuring spoon (optional)

❖ 4 medium glasses

INGREDIENTS

❖ 1 medium, very ripe banana, peeled

❖ 1 cup fresh strawberries

❖ 1 cup nonfat plain yogurt OR low-fat milk

❖ 1/4 teaspoon vanilla (optional)

❖ 2 ice cubes

Oliver says:

Dairy foods are the best source of calcium. That means strong bones, teeth, and muscles. One cup of milk or yogurt, 1 ounce of cheese, 1 1/3 cups of cottage cheese, or 1 cup of frozen yogurt equal a serving. Have 3 to 4 servings each day. Pick some low-fat or fat-free foods here whenever you can.

74

BREAK the banana into 3 or 4 pieces, and place in a blender container OR the workbowl of a food processor.

RINSE the strawberries, and pull off the stems. Put them in the container.

ADD the rest of the ingredients. Put the lid of the blender tightly in place.

75

Ask an Adult **BLEND** for about 30 seconds, or until smooth and creamy.

POUR into glasses and serve.

Mickey's Nutri Tip

I like to have this for a quick snack, or a good, fast breakfast. It's got calcium, vitamins A and C, and lots of fiber. Try other fruits like kiwi, pineapple, melon, or peaches for a new taste each time.

Desserts

- Chip 'n' Dale's Chocolate Peanut Clusters

- King Louie's Banana Boat

- Pinocchio's Pear Sundae

- 101 Dalmatians' Brownies

- Happy's Crunchy Rice Squares

- Snow White's Applesauce Cookie Bars

Chip 'n' Dale's Chocolate Peanut Clusters

Makes about 20 pieces

UTENSILS

* ❖ Medium microwave-safe bowl
* ❖ 1 cup dry measuring cup
* ❖ Oven mitt
* ❖ 1/2 cup dry measuring cup
* ❖ 1 mixing spoon
* ❖ 1 teaspoon measuring spoon
* ❖ 2 dinner plates

INGREDIENTS

* ❖ 1 cup reduced-fat semisweet chocolate chips
* ❖ 1/2 cup unsalted dry roasted peanuts
* ❖ 1/2 cup light OR dark raisins

POUR the chocolate chips into a microwave-safe bowl.

MICROWAVE on HIGH for 2 to 2 1/2 minutes, or until almost melted.

REMOVE the bowl from the microwave using an oven mitt.

ADD the peanuts and raisins to the chocolate and stir well.

DROP the mixture by teaspoonsful onto the plates.

REFRIGERATE for about 15 minutes, or until the clusters harden. Gently twist to remove from plates.

Ariel says:

There are no good foods or bad foods. Some foods you definitely want to eat more of than others. Eat plenty of beans, corn, apples, bananas, rice, lean meats and low-fat milk each day. Then sometimes you can have cake, chips, and candy, too.

79

Mickey's Nutri Tip

Peanuts and raisins are both high in fiber and carbohydrates. That makes them a high-energy food. Did you know that peanuts are also a good source of protein?

King Louie's Banana Boat

Serves 1

UTENSILS

- ❖ Toaster oven
- ❖ Cutting board
- ❖ Knife
- ❖ 8-inch piece of aluminum foil
- ❖ 1 tablespoon measuring spoon
- ❖ Oven mitt

INGREDIENTS

- ❖ 1 medium banana, peeled
- ❖ 1 tablespoon chocolate chips
- ❖ 1 tablespoon miniature marshmallows

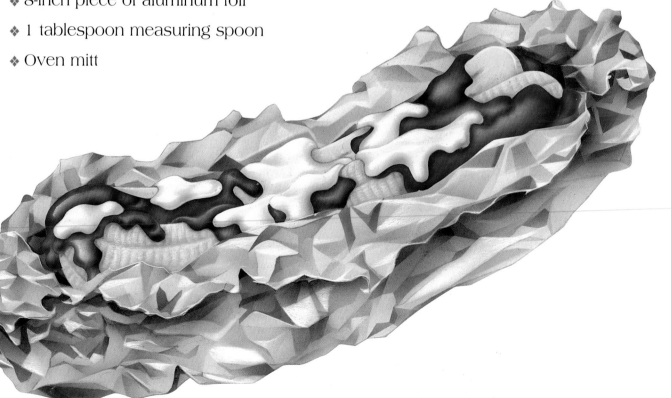

PREHEAT a toaster oven (or regular oven) to 400°F.

PLACE the banana on a cutting board, and cut it in half, making 2 long pieces.

PLACE the pieces side by side on the aluminum foil.

SPRINKLE the chocolate chips and marshmallows on top of the banana.

BRING up the sides of the foil to cover the banana and crimp, making a sealed package.

PLACE the banana package in the middle of the oven (directly on the rack), and bake for 15 to 20 minutes, until the chocolate chips and marshmallows are melted.

REMOVE the packet from oven using an oven mitt. Wait 1 minute for it to cool a bit.

UNFOLD the foil carefully. Fold down the sides to make a small boat.
King Louie's banana boat may be served right in the foil boat.

Mickey's Nutri Tip

This ape is no dummy. He knows how to take a little chocolate and make it a healthy snack. Do you think he knows bananas are a high-energy food? They are also a good source of potassium and carbohydrates.

Pinocchio's Pear Sundae

Serves 4

UTENSILS

❖ Can opener

❖ Medium strainer

❖ 2 medium bowls

❖ 4 dessert (small) plates

❖ Ice cream scoop

❖ 1 teaspoon measuring spoon

INGREDIENTS

❖ One 15-ounce can pear slices, packed in juice

❖ 4 sponge cake dessert shells*

❖ 2 cups low-fat frozen yogurt (your favorite flavor)

❖ 8 teaspoons chocolate syrup

*Look for the dessert shells in the deli or produce section of the supermarket. They are always available when fresh strawberries are in season. Otherwise, use slices of low-fat pound cake, or toasted whole wheat waffles.

OPEN the can of pears.

DRAIN the juice from the pears using a strainer, with a medium bowl underneath (to catch the juice).

PLACE one dessert shell on each plate. Place one scoop (about 1/2 cup) of yogurt on top.

ARRANGE about 4 pear slices (or enough to use all the pears) around the edge of each dessert.

DRIZZLE two teaspoons of the chocolate syrup over each plate of yogurt and pears, and serve.

83

101 Dalmatians' Brownies

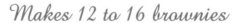

Makes 12 to 16 brownies

UTENSILS

- ❖ 8-inch square baking pan
- ❖ 2 medium mixing bowls
- ❖ Electric mixer
- ❖ 1/2 cup dry measuring cup
- ❖ 1 cup liquid measuring cup
- ❖ 1/3 cup dry measuring cup
- ❖ 1/4 cup measuring cup
- ❖ 1 teaspoon measuring spoon
- ❖ 1/2 teaspoon measuring spoon
- ❖ 1 tablespoon measuring spoon
- ❖ Mixing spoon
- ❖ Oven mitt

INGREDIENTS

- ❖ Nonstick cooking spray
- ❖ 1/2 cup sugar
- ❖ 1/3 cup vegetable oil
- ❖ 1/3 cup light corn syrup
- ❖ 1 teaspoon vanilla extract
- ❖ 2 egg whites
- ❖ 2/3 cup all-purpose flour
- ❖ 1/2 teaspoon baking powder
- ❖ 1/2 teaspoon salt
- ❖ 2 tablespoons unsweetened cocoa powder

Here's a good snack to have with a glass of cold milk. If you have made brownies before, you know these are different. They have less fat and no egg yolks. The chocolate-brown spots on the white batter are fun. And that's what good eating is all about.

PREHEAT the oven to 325°F (300°F if using glass). Lightly spray the baking pan with non-stick cooking spray. Set aside.

COMBINE the sugar, oil, corn syrup, vanilla, and egg whites in a mixing bowl. Beat on low speed for 1 to 2 minutes until well blended. Turn off mixer.

ADD the flour, baking powder, and salt. Beat again on low speed until just blended.

REMOVE 1/2 cup of the mixture. Put it into the other mixing bowl. Add the cocoa, and stir until blended.

POUR the white batter into the prepared baking pan. Spread with the back of a spoon to cover the bottom of the pan.

DROP the chocolate batter on top of the white batter using a teaspoon to make lots of spots. (Use your clean finger to push it off the spoon!)

 BAKE for 25 minutes, or until the brownies feel firm to the touch.

 REMOVE from the oven, using oven mitt. Let cool, and cut into squares.

85

Minnie Minder

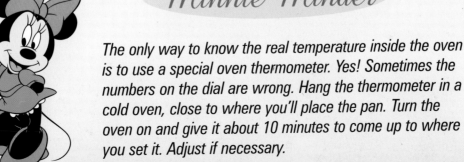

The only way to know the real temperature inside the oven is to use a special oven thermometer. Yes! Sometimes the numbers on the dial are wrong. Hang the thermometer in a cold oven, close to where you'll place the pan. Turn the oven on and give it about 10 minutes to come up to where you set it. Adjust if necessary.

Happy's Crunchy Rice Squares

Makes 24 squares

UTENSILS

- ❖ 13 x 9-inch baking pan
- ❖ Large mixing bowl
- ❖ 1 cup dry measuring cup
- ❖ 1/2 cup dry measuring cup
- ❖ Large mixing spoon
- ❖ Large microwave-safe bowl
- ❖ 1 tablespoon measuring spoon
- ❖ Oven mitt
- ❖ Knife

INGREDIENTS

- ❖ Nonstick cooking spray
- ❖ 5 1/2 cups crisp rice cereal
- ❖ One 6-ounce package dried chopped mixed fruits (about 1 cup)
- ❖ 1/4 cup toasted wheat germ
- ❖ 1 teaspoon ground cinnamon
- ❖ One 10-ounce bag large marshmallows
- ❖ 2 tablespoons vegetable oil
- ❖ 2 tablespoons low-fat milk

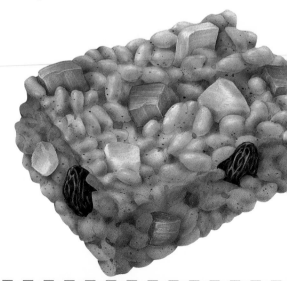

SPRAY the baking pan with nonstick cooking spray. Set aside.

COMBINE the cereal, dried fruit, wheat germ, and cinnamon in a large mixing bowl. Set aside.

COMBINE the marshmallows, oil, and milk in a large microwave-safe bowl. Microwave, uncovered, on HIGH for 1 minute. Stir the mixture. Microwave again, on HIGH, for 1 minute, or until the mixture is completely smooth.

REMOVE the bowl from the microwave, using the oven mitt. Pour the marshmallow mixture over the cereal. Stir to mix well. This gets thick, so you may need help here. (Some of the wheat germ may still be left in the bowl. Just stir in as much as you can.)

SPREAD the mixture into the baking pan using a large mixing spoon, and make an even layer. (You can also spray some cooking oil on your fingers—washed, please—to pat the mixture into the pan.)

Ask an Adult **PUT** the pan into the refrigerator and chill until just firm. Cut into squares.

87

Snow White's Applesauce Cookie Bars

Makes 24 bars

UTENSILS

- ❖ 13 x 9-inch baking pan
- ❖ Large mixing bowl
- ❖ 1 cup dry measuring cup
- ❖ 1/2 cup dry measuring cup
- ❖ 1 teaspoon measuring spoon
- ❖ 1/4 teaspoon measuring spoon
- ❖ Whisk
- ❖ Medium mixing bowl
- ❖ 1 cup liquid measuring cup
- ❖ Electric mixer
- ❖ Rubber scraper OR spoon
- ❖ Oven mitt
- ❖ Toothpick OR cake tester
- ❖ Wire cooling rack
- ❖ Knife

INGREDIENTS

- ❖ Nonstick cooking spray
- ❖ 1 1/4 cups all-purpose flour
- ❖ 1/2 cup whole wheat flour
- ❖ 2 teaspoons ground cinnamon
- ❖ 1 teaspoon baking powder
- ❖ 1/4 teaspoon ground nutmeg
- ❖ 1/4 teaspoon salt
- ❖ 1 cup unsweetened applesauce*
- ❖ 1/3 cup vegetable oil
- ❖ 1 cup packed light brown sugar
- ❖ 1 cup dark raisins

*If sweetened applesauce is all you have, use the same amount but reduce the light brown sugar to 3/4 cup.

PREHEAT oven to 350°F. Spray the baking pan with nonstick cooking spray. Set aside.

WHISK or stir together the two flours, cinnamon, baking powder, nutmeg, and salt in a large bowl.

COMBINE the applesauce, oil, brown sugar, and raisins in a medium bowl. Use an electric mixer, on low speed, and beat for 1 minute, or until the brown sugar is dissolved.

POUR the applesauce mixture into the flour mixture. Stir in the raisins. Beat again on low speed, just until blended.

SCRAPE the batter into the pan, using a rubber scraper or spoon. Spread to make even layer.

PLACE the pan in the oven, using oven mitt. Bake for about 25 minutes. Use a toothpick (or cake tester) to poke the middle of the pan. If it comes out dry, the bars are done. If not, bake a few minutes longer.

TAKE the pan out of the oven, using oven mitt. Place it on a wire rack to cool. Cut into squares.

Minnie Minder

You can also use a 12 1/4 x 8 1/4-inch foil pan for this recipe (like Happy's Crunchy Rice Squares, page 86). It's an easy way to take these bars to a party, or to school. Don't forget to recycle!

Putting It All Together

Here are some ways for you to take some of the recipes and combine them into a meal or a snack.

HOLIDAY BREAKFAST

Pineapple or Cranberry Juice Cocktail
Mickey's Whole Wheat Honey Pancakes (page 20)
Sliced Banana
Low-Fat Milk

EVERYDAY BREAKFAST

Orange Juice
Goofy's Smart-Start Oatmeal (page 16)
Low-Fat Milk

INVITE A FRIEND FOR LUNCH

Goofy's Easy as A-B-C Vegetable Soup (page 30)
Whole Wheat Roll
Peter Pan's Sparkling Pink Punch (page 72)
Graham Crackers

WHAT'S FOR DINNER?

Tossed Salad with Low-Fat Dressing
Tramp's Favorite Pasta (page 56)
Italian Bread

Tossed Salad

Use a big bowl and make sure all the veggies are washed and dried before you use them. Use several of the items below each time you make a salad. Once you have the veggies in the bowl, add a few tablespoons of low-fat dressing or sprinkle a little oil and vinegar on top. Then toss well, using two large spoons or salad tongs. (To toss a salad means gently combine all the ingredients within the bowl. No veggies in the air, please!)

- *romaine lettuce*
- *iceberg lettuce*
- *cherry tomatoes*
- *cucumber slices*
- *mushroom slices*

- *radish slices*
- *avocado cubes*
- *red or green pepper strips*
- *red or white canned beans (drained)*

Baked Potato

Preheat oven to 350°F. Scrub four medium-size potatoes, and pat dry with a paper towel. Pierce the pototoes a few times with a fork. Bake for about 45 minutes, or until tender.

WEEKEND DINNER

Tomato Juice Appetizer
Mufasa's Mouthwatering Chops (page 66)
Thumper's Buttered Baby Carrots (page 46)
Baked Red Potatoes
Frozen Yogurt

AFTER SCHOOL SNACKS

Drink a glass of low-fat milk or chocolate milk with any one of these fun treats:

Aladdin's Magic Carpet Rolls (page 26)
Rafiki's Coconut Fruit Kabobs (page 42)
Happy's Crunchy Rice Squares (page 86)
Abu's Baked Apples (page 40)

WE'RE HAVING A DINNER PARTY

The Sultan's Hummus Dip with Veggies (page 38)
Hercules' Powerhouse Meatloaf (page 60)
Peas and Carrots
Alice's Cheesy Rice (page 48)
Pinocchio's Pear Sundae (page 82)
Mrs. Potts's Spiced Cider (page 70)

The Perfect Setting

A table that's pretty and clean makes the food taste extra good!

SETTING THE TABLE

❖ Dust or wipe the table with a damp—not wet—sponge.

❖ Use a tablecloth to cover the table or placemats for each person.

❖ Place a plate or bowl for each person about 1 inch from the edge of the table.

❖ Place a knife by the right side of the plate. Turn the blade of the knife toward the plate.

❖ Place a spoon next to the knife.

❖ Place a folded napkin on the left side of the plate.

❖ Place the fork by the left side of the plate (on top of the napkin, if you're not using napkin rings) with the tines (the part that holds the food) facing up.

❖ Place the beverage glass on the right, just above the tip of the knife.

❖ Place the salad plate or bowl (if you are having one) on the left above the fork.

CENTERPIECES

It's nice to have a centerpiece—a pretty arrangement in the middle of the table—especially for parties or special occasions. You can make your own centerpiece with lots of items found around the house or yard—the key is to be creative and have fun! Try fresh flowers in a vase, or make cut-out flowers from colored paper attached to stems made of pipe cleaners or sticks. For a theme, use holiday ornaments or even a favorite toy collection of bears or dolls. A bowl of fresh fruit, pine cones, or even colored balls, look terrific, too. Check out your library for books on creating imaginative and colorful centerpieces for every occasion. There are only two rules: Leave room for the plates and food! And don't create anything so high that you can't see your friends and family across the table!

NAPKINS

everyone needs to stay tidy at the table, so napkins are a ust! Whether paper or cloth, napkins can help adorn a ble before they are placed in your lap. Here are some asy and fun ways to use napkins while setting the table.

❖ Fold a napkin into a triangle and place beneath the fork.

❖ Use napkin rings—store-bought ones come in a variety of shapes, colors, and themes. Or you can make your own! Have an adult help you cut a paper towel roll into 1 1/2 inch strips (be careful not to crush the roll). Paint each strip with zany

designs, or glue on macaroni in different shapes, or add glitter. You can also cover the strips with colored paper or felt. Each place at the table can have a different design. Mix and match, use your imagination. Once the rings are done (and make sure they're dry), roll the napkins up and poke through the rings. Place them at the top of the plate or directly on it.

❖ Fold or roll the napkin so that it is an inch or so wide, then wrap it around the middle with different colored yarns or ribbons, or tie it with a bow. Center on the plate.

Index

To Marlene and Jerry, Danielle and Rachael;
with much love and many thanks.

First Disney Press Paperback Edition 1999
Recipes © 1998 by Pat Baird.
Recipe Titles © 1998 Disney Enterprises, Inc.
Additional Texts © 1998 Disney Enterprises, Inc.
Artwork © 1998 Disney Enterprises, Inc.
Compilation © 1998 Disney Enterprises, Inc.

Printed in Singapore.

First Edition
1 3 5 7 9 10 8 6 4 2

Designed by Atif Toor.
Food Illustrations by Cindy Sass.

Library of Congress Catalog Card Number: 97-80313
ISBN: 0-7868-3230-4 (paperback)

For more Disney Press fun, visit www.disneybooks.com